The Emperor's
New Clothes

The Emperor's New Clothes

Biological Theories of Race at the Millennium

JOSEPH L. GRAVES JR.

RUTGERS UNIVERSITY PRESS

New Brunswick, New Jersey, and London

To the memory of my brother, Warren Graves, M.D.
(December 14, 1957–March 6, 1998),
and all those who suffered through a racist world
not of their own making.

Also to my sons, Joey and Xavier,
and the children of the twenty-first century.
May you grow up to a world that has learned
that there are no races and that we judge people
by the content of their character.

Third paperback printing 2008

Library of Congress Cataloging-in-Publication Data
Graves, Joseph L., 1955–
    The Emperor's new clothes: biological theories of race at the millennium /
Joseph L. Graves Jr.
      p. cm.
    Includes bibliographical references (p.).
    ISBN 0-8135-3302-3 (alk. paper)
      1. Race. 2. Physical anthropology. 3. Sociobiology. 4. Racism in
anthropology. I. Title.

  GN269.G73 2001
  599.97—dc21   00-034205

British Cataloging-in-Publication data for this book is available from the
British Library.

Manufactured in the United States of America

Contents

Preface to the 2005 Printing

The Emperor's New Clothes was written between the spring of 1998 and fall of 1999, and after much revision, the hard cover edition was published two years later in 2001. The book summarized conceptions of race generated from physical traits and protein polymorphisms. Over the last four years, research on human genetic variability, particularly on its implications for medicine, has greatly expanded. In 2002, the draft human genome sequence was published. As a result, Celera Genomics CEO, J. Craig Ventor, stated that molecular data had confirmed that "race is a social concept." Actually, the human genome project could not evaluate the nonexistence of human races, since it was never designed to examine population-based DNA variation. It was supposed to be a map of the linear sequence of nucleotides in the human genome, nothing more. However, scientists involved in the project already knew that the distribution of human genetic variation invalidated the classical conceptions of race.

The interpretation of new studies of human genetic variation took center stage at a symposium entitled "Human Genome Variation and Race," held at Howard University in the spring of 2003, which I attended as a member of the external advisory board of Howard's National Human Genome Center.[1] There was general agreement that socially-constructed and biological theories of race do not coincide. However, there was significant disagreement about whether, at some level, genetic variation in humans can be clustered into groups based on geographic region.

Throughout the discussions, it became clear that many of the participants' views were limited by their lack of familiarity with the history and philosophy of racial classification. Ironically, those responsible for generating new and critically important data on human genetic variation knew little of the path they had been following. As a potential solution to this contradiction, *The Emperor's New Clothes* remains in the forefront, providing the reader with both a summary and analysis of some of the great racial questions posed by the most important thinkers in biology through the eighteenth to the twentieth centuries.

The Emperor's New Clothes weighs in heavily against arguments that new molecular data justifies dividing humans into racial groups.[2] For example, the vast majority of human DNA (85 percent) is never coded to RNA and of the 15 percent coded only a fraction (one-fifteenth) produces the proteins that run our cells. For this reason, more genetic variation is maintained in the noncoding fraction.

Genetic changes in the coded DNA can be acted on by natural selection. But the coding dictionary is redundant when changes in a nucleotide base Adenosine (A), Thymine (T), Guanine (G), Cytosine (C) lead to the same amino acid; in these cases the function of the resultant protein is unaltered. Thus natural selection will not be important in altering the frequency of such changes. However, in cases were the nucleotide change leads to a new amino acid, particularly an amino acid that has a different charge, the resultant protein's function could be altered. Since most of these alterations are likely to be bad for reproductive fitness, natural selection has a purifying effect on these mutations by reducing their frequency. In the rare case that the mutation leads to better function, natural selection is likely to increase the frequency of such successful variations.

The greater genetic variability found in the noncoding sections of DNA makes it useful for forensic analysis. For example, short tandem repeat (STR) sequences can be used to uniquely identify individuals. These alleles are two- to five-base-pairs-long and are dispersed throughout the human genome. Since there is a high variety of alleles at each locus, they distinguish individuals well. These portions of DNA are inherited, so children inherit the alleles of their parents. Thus relatives share more genes in common than nonrelatives. In addition, people in the same geographic region are more likely to be related to each other and the same sort of forensic analysis that distinguishes individuals and families can be employed to distinguish populations on a regional basis. In this way, noncoding DNA can provide a much stronger clue to the ancestry of individuals. This works as a simple function of the power rule of probability. If two genetic sequences are inherited independently, the chance of them being shared by any two individuals is the product of their frequency. Thus, if two individuals have a 0.095 probability of sharing the same genotype at one locus, they only have a $(0.095) \times (0.095) = 0.00902$ probability at two loci. The product for six loci would be $(0.095)^6 = 0.000000735$, or 7.35 out of ten million times would they have the same genotype. Each STR genotype is rare, so the rate at which individuals fail to match increases rapidly with the addition of another statistically independent locus, (those found on different chromosomes).

This analysis allows the clustering of individuals from various regions in the world, but the clustering on noncoding DNA segments does not guarantee that individuals can also be unambiguously clustered using coding segments of DNA, or separated into "races." The coding segments produce physical features that have been acted on by natural selection in a discordant fashion (see Chapter 9.) Thus, genes that code for features such as skin color are not necessarily correlated with those associated with body proportions, skull characteristics, disease predisposition, or intelligence. A summary of my explanation of this problem can be found at a Web forum entitled Race and Genomics, http://raceandgenomics.ssrc.org, sponsored by the Social Science Research Council, posted in April 2005.[3]

Race and Medicine

Chapter 11, "The Race and Disease Fallacy," retains all of its salience to recent controversies in race and medicine. For example, the new drug BiDil has been demonstrated to reduce the death rate from congestive heart failure in African Americans by 43 percent. It was designed and tested to improve health outcomes in a specific population, thus erroneously dubbed "the racial pill" in the popular media.[4] BiDil is a combination of a nitric oxide donor isosorbide dinitrate and the antioxidant hydrazaline, which also acts as a vasodilator. Nitric oxide is a gas that plays a role in a variety of neurally mediated events including regulating heart processes, in programmed cell death, as an antimicrobial agent, and even in penile erection in men. Antioxidants protect cells against oxidative damage that results from normal cellular respiration and poisons that accumulate over time.

The researchers involved in the African American Heart Failure Trial (A-HeFT) were motivated to try this combination of drugs by studies that showed that people who self-identified as "black" had lower levels of available nitric oxide and greater amounts of oxidative stress than those who self-identified as "white."[5] It has been demonstrated that oxidative damage to human cells results from periods of prolonged stress.[6]

Despite the rhetoric, BiDil is not a "racial" pill. We know that nitric oxide is synthesized by individual cells and this is catalyzed by an enzyme known as endothelial nitric oxide synthase (eNOS). Researchers in the Bogalusa Heart Study found that there is genetic variation at position G894T in this enzyme that influences arterial stiffness (after controlling for sex, age, body mass index, insulin, heart rate, and mean arterial pressure).[7] African Americans who had the T allele had less elasticity than those with the G allele. European Americans with the T allele also showed less elasticity, but the difference against those with the G allele was not significant for the number of people tested. They also found that the frequency of T was 0.131 in African Americans versus 0.321 in European Americans. Thus European Americans have a higher frequency of alleles that should predispose them to heart disease. Some other factors must be reducing the negative impact of this allele in "whites" that is not operating in "blacks." Yet once again, the data do not support this, meaning that other factors must be at play. Throughout this book I explain why and how social dominance creates different environmental conditions for the socially constructed races in America. New studies have now demonstrated molecular mechanisms by which this dominance leads to cellular damage.[8]

The second printing of the paperback edition comes at a time when clarity on the social construction of race is needed more than ever. Recent advances in genetic research raise new questions. Their solution requires that we understand the history of racial reasoning in biology. In addition, the debate on social justice is revitalized, particularly in the light of the last election, the war in Iraq, and the right-wing assault on the courts and science in the public schools. What follows is crucial for grasping the meaning of these events in the context of the historical debate on race.

Dr. Joseph L. Graves, Jr.
April 27, 2005

Acknowledgments

First, I would like to acknowledge my wife, Suekyung, without whose support I could never have written this book. In addition, I would like to thank the students at Arizona State University and the University of California, Irvine, who enrolled in the courses that generated much of the material that became this book. Special thanks must also be extended to the following individuals, who directly or indirectly supported my efforts to write this book: Denise Green, my senior office assistant; and Jessie Durbin, Thao Pham, and Shannon Hoss, my research laboratory technicians. Margaret Hayes helped with the initial library research. I would also like to thank my colleagues at both campuses whose discussions contributed to this work.

The Emperor's
New Clothes

Racial Thinking
Complaints and Disorders

The trumpets blew and the great procession started. People had come from far and near to get a good view of the Emperor in his new clothes. But how surprised they were when they saw him. At last one of the crowd said timidly, "The Emperor's new clothes are beautiful!" Then everyone started talking at once. "So fashionable!" "Very smart!" "Divine!" they said, each of them anxious not to seem more foolish than the rest.

But one small boy laughed out loud and shouted, "Look! The Emperor has no clothes on!"

Hans Christian Andersen, *The Emperor's New Clothes*, 1837

The story of the emperor's new clothes has become the time-proven metaphor for patently false theories. In the case of biological theories of "race," however, the analogy might seem somewhat strained. But it is not. The honesty and naivete of the child in Hans Christian Andersen's fairy tale are sufficient to demonstrate the foolishness of the emperor and his public. The child is capable of honest observation and is willing to state the obvious. Dismantling biological theories of race requires no more than the same will. However, Andersen's fairy tale does not communicate the dire need to eliminate racist ideology. In his tale, the emperor is simply vain and foolish. Vanity and foolishness, though indeed problems, are not as grave as racism. Racism is more than foolish; it is evil and destructive. I have written this book because I believe that our society cannot progress toward true justice and equality until we exorcise racism from our collective consciousness.

Specifically, my goal is to show the reader that there is no biological basis for separation of human beings into races and that the idea of race is a relatively recent social and political construction. To accomplish this goal, I examine the history of theories of biological diversity from a modern scientific perspective. Because racial theories have always been interwoven with social and political policy, this work is by necessity interdisciplinary.

Demolishing the idea of biological race lays bare the fallacies of racism. If biological races do not exist, then what we call "race" is the invention not of nature but of our social institutions and practices. The social nature of racial categories is significant because social practice can be altered far more readily than can genetic constitution. Although professional scientists generally agree on the fallacy of race, this idea has yet to find its way into the public discourse.

Many nonhuman species do have subspecies levels of genetic variation. Indeed, the existence of geographic races (subspecies) is essential to the concept of organic evolution. So there is nothing that precludes the existence of races in sapient beings. In fact, early hominids might have been classifiable into valid racial categories, in which, for example, pairs of individuals from different races either had reduced capacity, or no capacity, to produce viable offspring. If this were the case for modern humans, then racist thinking might find some scientific support, but the simple fact is that these categories do not exist in modern humans.

Thus one does not have to compromise one's scientific integrity to be morally opposed to racism and other forms of bigotry. My personal experiences with racism as an African American intellectual have certainly given me perspective on the harm caused by racist thinking and practice. However, even if racism were not morally objectionable on the basis of the harm it causes, objective science would still disprove the existence of races in the human species. In the absence of a biological basis for race, racism simply becomes ideology. As ideology, it is rightly subject to moral judgment. For this reason, people wedded to racist ideology will object to this work and its approach, because it denies them the scientific high ground. Racist ideologues have been accustomed to the luxury of hiding behind so-called reasoned objective argument while characterizing their critics as emotional or "politically correct." By demonstrating that racist science is critically flawed, we lay bare their hidden agenda.

However, I address this book not to racist ideologues but to people who are confused and unsure about what we mean by race and who feel intuitively that racism is wrong. I believe that the vast majority of people fall into this latter group. Like the emperor in Andersen's fable, the "emperor race" has paraded through history naked for everyone to see. All we need do is observe biological diversity carefully, report it faithfully, and not fear the consequences of the truth that is revealed.

Some Thoughts on Race in History

Race and racism were fundamental forces in the founding of the United States of America. Many of our present political and social problems stem from that fact. The tragedy is that modern biology allows us to recognize that the concept of race is fundamentally flawed. However, the eighteenth-century "Enlightenment" scholars never doubted that God and science declared the existence of races and that there should be hierarchical relations among them. According to this thinking, the European stood at the pinnacle of human perfection, and all other races were to be measured against him.

Clearly, we have come a long way since that period. Many assertions and assumptions about race and racial relations that were taken for granted during the Enlightenment have subsequently been proven false (such as the incorrect assertion that Negroes' brains are smaller than those of white Europeans). This change in thinking did not happen without tremendous struggle; the ideological battle against racism has now been fought across three centuries. Nevertheless, people continue to suffer and die as a consequence of racist policy. Even today the root cause of racism remains entrenched in the American consciousness. Many Americans still believe that there are innate racial differences among people in character and habit.[1]

However, humans did not always feel this way; race theory is a consequence of relatively modern historical developments. Clearly the ancients recognized that human beings were physically different from one another and that they had formed different cultures. However, Western civilization did not immediately develop substantial ideological support for theories of race classification and racially based variation in character and temperament. The rise of racial ideology coincided with the development of social institutions that exploited human biological difference for profit.

What I am attempting in this work will not be easy for many people to digest. Race is part of the American legacy, and racial exploitation gave the United States license to exist. Many people still unconsciously revel in the European pioneer spirit of discovery and domination. Europeans have been portrayed in the context of superiority to the "other," with the special qualities of their race explaining their power. Although many think it is of little consequence, Thomas Jefferson and some other signatories of the Declaration of Independence were slaveholders. Even many who were not slaveholders had fortunes that could be linked to the institution of chattel slavery. It must be remembered that all the colonies were founded in regions that had formerly been the sole dominion of the "native" populations of this continent. These "native" people were themselves descended from migrants, most likely originating in eastern Asia. Colonization by Europeans drove many of these people off their tribal lands and established "Caucasian" sovereignty. Much of the Southwest became part of the United States as a result of a war against Mexico meant to extend the territory of the slaveholders.

Blacks and American Indians are not the only people who have been subject to racism in this country. Few consider the origin of expressions like "not a Chinaman's chance" or recognize the sacrifice that Chinese Americans made to extend the railroads in the American West. Many never questioned the dubious logic of herding Japanese Americans into concentration camps during World War II while allowing German Americans to go free. Even the Irish and the Italians did not become "white" until well into the twentieth century, and anti-Semitism has been a consistent backdrop to American history.

Throughout our history, Anglo-Saxons and other northwestern European populations have often enjoyed prosperity at the expense of other groups. This prosperity has been attributed to the biological superiority of the Anglo-Saxon and Teutonic races. Unfortunately, very few Americans recognize the true reasons for this selective prosperity, the sacrifices that all the immigrants to this continent (that means all of us) made to create this modern nation.

In the main, American scholars of the last three centuries have strained to justify the special position of the Euro-American populations in our democracy. The scholars' theology told us that the supremacy of European Christians had been ordained by God. The inferior peoples of the earth had been provided to the Christians by way of divine Providence. Pre-Darwinian biology invoked the *scalae naturae*, the "great chain of being," in which humans ranked higher than all other life forms; and later, Darwin's theory would be co-opted to demonstrate that the biological characteristics of Europeans were responsible for their world dominance. The nineteenth century would spawn new pseudoscientific theories related to race and human biological diversity, among these social Darwinism and eugenics, the study of human "improvement" by genetic control. Anthropology in the nineteenth century was almost wholly devoted to the research paradigm of Anglo-Saxon, Teutonic superiority. Because the modern research land-grant university did not yet exist, only a few people, mainly well-to-do Euro-American men, had access to careers in science. For this reason, alternative voices were marginalized; women and "people of color" generally could not earn professional degrees in scientific disciplines.

The early twentieth century provided the scientific means by which much of race theory could be tested. Early experiments in genetics began to demonstrate the underlying complexity of human biological characteristics. Anthropological investigations of human blood groups returned results inconsistent with nineteenth-century racial classification, which had assumed that blood groups differed between the races. Eugenics, the vogue of the first half of the twentieth century, would be both politically and scientifically discredited by the 1950s. The birth of molecular biology in the 1940s and its exponential growth in the latter half of the century provided the final tool for the dismantling of the race concept. Today, the majority of geneticists, evolutionary biologists, and anthropologists agree that there are no biological races in the human species.[2] We have known this for at least a decade; yet the message has not been successfully conveyed to the American public.

Race and Racism: Basic Definitions

The term "race" implies the existence of some nontrivial underlying hereditary features shared by a group of people and not present in other groups. Biological science has long been interested in the identification and quantification of variation within species and has developed relatively precise tools to examine the hereditary characteristics exhibited by organisms. These developments were instrumental in allowing the Western, socially constructed concept of race and the biological concepts of race to diverge. None of the physical features by which we have historically defined human races—skin color, hair type, body stature, blood groups, disease prevalence—unambiguously corresponds to the racial groups that we have constructed.

Physical traits do vary among geographical populations, although not in the ways most people believe. For example, Sri Lankans of the Indian subcontinent, Nigerians, and Australoids share a dark skin tone but differ in hair type and in genetic predisposition to disease. Further difficulty results from the fact that people commonly link directly observable physical variation with less directly observable variation in such putative attributes as intelligence, motivation, and morality. However, the lay concept of race does not correspond to the variation that exists in nature. Instead, the American concept of race is a social construction, resulting from the unique political and cultural history of the United States.

Sadly, the lay public cannot necessarily turn to a dictionary or encyclopedia for a reasonable definition of race. For example, the *New Merriam-Webster Dictionary* (1989) gives two definitions of "race": "family, tribe, people or nation of the same stock; also: mankind" and "a group of individuals within a biological species able to breed together." Both definitions are misleading, and the second is a tautology. A species is defined precisely as a group that is capable of interbreeding. The *American Heritage Dictionary* (3d ed., 1992) contains the following definitions of "race":

1. A local geographic or global human population distinguished as a more or less distinct group by genetically transmitted physical characteristics. **2.** A group of people united or classified together on the basis of common history, nationality, or geographic distribution: *the German race.* **3.** A genealogical line; a lineage. **4.** Human beings considered as a group. **5.** *Biology.* **a.** A population of organisms differing from others of the same species in the frequency of hereditary traits; a subspecies. **b.** A breed or strain, as of domestic animals. **6.** A distinguishing or characteristic quality, such as the flavor of a wine. [French, from Old French, from Old Italian *razza,* race, lineage.]

Only number five is a correct scientific definition of race. When we apply this definition to the human species, we readily arrive at the conclusion that no biologically defined races exist in the human species. That is, if race is defined as a population that has achieved the subspecies level of genetic differentiation, no such divergence currently exists in our species. It is the confusion between definitions one and five that creates the controversy. How much genetic difference must there be before a subspecies can be said to exist? Do measured differences have anything to do with the nature of human society and, if so, how? It is also important to recognize that definition number five has only been available since the advent of the neo-Darwinian revolution, which began at the turn of the twentieth century and culminated in the 1950s.

Microsoft's *Encarta 99* gives a similarly inadequate definition of the "human race": "The identification within a species of subpopulations whose members share with one another a greater degree of common inheritance than they share with individuals from other such subpopulations. The primary application of the concept of race is to subpopulations of the human species, and race is thus a term that ordinarily applies to groups of people. Applied to an individual, race refers to membership in a group, and not to aspects of the person's appearance, such as skin color." This definition is self-contradictory. Although it states that a race is a population that differs from other populations by some unspecified degree of common inheritance, it also states that race is not a sufficient category with which to define a given individual's physical traits. Later in the definition, examples of the existing five "races" of man are listed: Europeans and similar peoples, East Asians, American peoples, sub-Saharan African peoples, and finally South Asian, Australian, and Oceanian peoples. Each example is shown with a type specimen: a young Englishman, a young Miao (Hmong) girl, an elderly Lakota woman, a young woman from south of the Sahara, and a wigman from Papua, New Guinea. Only one picture is provided for each group; thus there is no sense of the wide variation of features found in each of these categories. These groups are in turn derived from the following classical anthropological categories, which are usually listed in the following order: Caucasian, Mongolian, American Indian, African, and Australoid. Even if these were valid "racial" categories, there would be no logical or evolutionarily valid reason for listing the populations in this order. The list is certainly not alphabetical; neither does it follow the accepted evolutionary relationships among these groups.

To understand the order and format in which these examples are presented, we must understand the history of racial anthropology. The examples are presented in precisely the order given in the racial taxonomy of eighteenth-century anthropologist Johann Friedrich Blumenbach. Blumenbach's scheme for the human races was mired in the typological biology of that period. Thus, *Encarta 99*'s definition of race has not really progressed from that which we could have found in the nineteenth century. No solution to this problem is provided by the *Encarta Africana 99*, which contains no biological definition of race.

From these examples, it is safe to assume that the American public still has little access to well-formulated and scientifically precise definitions of the concept of race. This situation results both from the general problem of conveying complex scientific concepts to the public through popular resources and media and also from the long-standing confusion among reputed scholars on the concept itself.

Human cultures did not begin with fully formed scientific communities. The development of the scientific method, and of communities of professional scientists who practice it, is really a phenomenon of the late nineteenth and twentieth centuries. Before there were scientific examinations of human diversity, unscientific definitions must by necessity have prevailed. Such definitions were by nature arbitrary, therefore inconsistent; and hypotheses concerning them would have been untestable. Even supposedly scientific examinations of human diversity were seriously flawed because they were not based on evolutionary theory. Thus it is not surprising that the previously existing scholarship on race was so often capable of drawing diametrically opposed and logically inconsistent conclusions. The tragedy, of course, is that for most of the history of human civilization, these conclusions found their way into the thinking of ordinary people in daily life, with devastating consequences.

What Is Racism? How Will I Know It?

Modern racist ideology wishes to appear as a part of normal intellectual discourse. Even worse, it attempts to portray its critics as the racists. Is the recognition of biological variation in humans necessarily racist? Is the attempt to classify such variation and show correlation between physical and mental characteristics necessarily racist? The answers to these questions very much depend upon how these processes are accomplished. For example, *Webster's* defines racism as "a doctrine without scientific support, that claims to find racial differences in character, intelligence, etc., that asserts the superiority of one race over others and that seeks to maintain the supposed purity of a race."

By this definition any view of a particular race that is not scientifically substantiated could be considered "racist." There are limitations in even this definition, but it shall serve our purpose for now. In *The Bell Curve* (1994), Richard J. Herrnstein and Charles Murray claimed that they had summarized more than a century's worth of scientific data substantiating their view of the statistically significant and repeatable IQ superiority of East Asians and Caucasians over Negroids. In their study, they repeatedly confused ethnicity and race. Taken at face value, such a study is not necessarily racist. In other words, examining, measuring, and quantifying human biological diversity does not necessarily make one a racist. However, I shall endeavor to demonstrate in this volume that Western attempts to do so have historically been both motivated by racist social agendas and infused throughout with racist ideology.

Herrnstein and Murray's study illustrates the following common errors, which are consistently associated with studies purporting correlation between race and complex behavior:

- making claims that are not substantiated by the data,
- making mathematical and statistical errors that conveniently support one's hypothesis,
- ignoring alternative hypotheses that can explain the results equally well,
- ignoring data inconsistent with one's own hypothesis,
- ignoring theory and data that challenge core assumptions, and
- making sweeping policy recommendations that conform to racist philosophies and practices.

I will show throughout this work that the history of the study of human variation in the United States coincides with *Webster's* definition of racism. What is also clear is that because many people (scholars included) assume the validity of our racial categories, they ascribe to them the power to explain human characteristics that they simply do not have. We must ask how this happens. Research indicates that although young children are aware of phenotypic differences (such as skin color) between individuals, they do not see these differences as significant.[3] However, most American adults assume that these same perceived differences predict the behavioral, intellectual, moral, and ultimately social worth of individuals.

Herrnstein and Murray assert in *The Bell Curve* that much social dysfunction is directly attributable to lower intelligence. They demonstrate a consistent racial difference in IQ and assert that this difference is genetic and highly heritable. According to them, the genetic character of intelligence means that inequities cannot be fundamentally altered by environmental interventions such as social programs (for example, affirmative action). They further argue that this documented difference in intelligence between groups predicts disproportionate achievement for individuals within those groups. Thus, the underrepresentation of African Americans (and other low IQ populations) on the higher rungs of the social ladder is not the result of historical discrimination but the natural result of free competition. In such competition, the most-qualified individuals supposedly succeed. Finally, Herrnstein and Murray argue that our present social system encourages lower IQ individuals to reproduce faster than higher IQ individuals. This, they predict, will cause the overall decline of intelligence in Western society as more low IQ genes are passed on to future generations.[4] None of this can be considered racist because it is based on sound scientific reasoning. Or is it? In parts 3 and 4 we shall carefully examine these arguments.

Why Should We Resist Racism?

Some disparage the argument that we should resist racism as an example of political correctness. It is not. The simple fact, as will be demonstrated in this book, is that science identifies no races in the human species, not because we wish there to be no races but because the peculiar evolutionary history of our species has not led to their formation. There is more genetic variability in one tribe of East African chimpanzees than in the entire human species![5] Only political orthodoxy in a racially stratified society has maintained the race concept for this long. If race does not exist at the biological level, then its use in social and political policy is profoundly flawed. Indeed, it is a falsehood in the service of social oppression.

Support for social oppression can be pervasive. Racist ideologies provide a moral justification for maintaining a society that routinely deprives certain groups of their rights and privileges. Racist beliefs discourage subordinated people from questioning their lowly status; to question that status is to question the very foundations of the society. In addition, racism focuses vague social uncertainty on a specific threat. Racism therefore not only justifies existing practices but also serves as a rallying point for social movements, such as fascism in Europe and the white supremacist movement in the United States. Finally, racist myths encourage support for the existing order. Some argue that any major societal change would lead to even greater poverty in subordinated groups and would lower the living standard of the dominant group. History demonstrates that the virulence of racism increases when a value system is under attack.

However, nothing comes without cost. There are at least seven ways in which racism and discrimination lead to dysfunction in a society, even for a dominant group:[6]

- Discriminatory practices prevent society from making use of the contributions of all individuals. Discrimination limits the search for talent and leadership to the dominant group. Racists may also view "inferior" people as a physical resource to be apportioned in the social division of labor for the benefit of the "superior" race.
- Discrimination aggravates social problems such as poverty, delinquency, and crime and places the financial burden of alleviating these problems on the dominant group.
- Racism requires society to invest a good deal of time and money to defend the social and institutional barriers that prevent the full participation of all members.
- Racial prejudice and discrimination undercut diplomatic relations among nations. They also negatively affect efforts to increase global trade.
- Racism restricts communication among groups. Little accurate knowledge of minorities and their culture is available to the society at large.
- Racism inhibits social change because change may assist a subordinate group.
- Discrimination promotes disrespect for law enforcement and for the peaceful settlement of disputes.

As I shall demonstrate in part 4, there are also scientific and medical costs to society that result from racist ideology. The aim of *The Emperor's New Clothes* is to lay bare the ideological supports for racism. The twisted world of race and racism undermines the social and political foundations of our society. Yet this entire world is based on a spurious fiction, a fiction that is utterly naked. We might gain enormous benefits if we exorcised racist ideology. I believe that the survival of the United States as a democracy depends on the dismantling of the race concept. This dismantling requires careful examination of the tortured history of race and racism.

PART 1

The Origin of the Race Concept

The history of biology can be separated into pre- and post-Darwinian periods on the basis of the rationale that evolutionary theory is required to make sense of biological phenomena. As a biological concept, race can be demarcated into these same periods. In the pre-Darwinian period, there is an additional dividing line of significance: theories about human diversity held before the Age of Discovery differ from those held afterward.

Before the fifteenth century, when human populations were limited in their ability to disperse, there were few recorded theories or ideas concerning what we think of today as race. On balance, the ideas that did surface at that time did not uniformly assign superiority to any particular race. People tended to be ethnocentric, but only in a very parochial sense. The "superior" people might have been those of a tribe or city. Eventually, the development of world travel facilitated by better navigation and sailing technologies allowed previously disconnected human populations to intermingle. The greater contact among world populations increased speculation concerning the nature and significance of human biological diversity.

In the pre-Darwinian period after the dawn of the Age of Discovery, religion and science were often in agreement concerning the fundamental nature of living organisms. Before Darwin, all theories of biology were fundamentally creationist and thus imputed some type of design to all the features of living things. Race, too, must have resulted from some supernatural act, the thinking went, and in some way must have had a designed significance. In the West, the rise of colonialism brought in a racial hierarchy that was linked to these religious and biological theories. Europeans, it was thought, had to be closest to God, and natural science was expected to confirm this preordained fact.

Darwin's publication of *The Origin of Species* (1859) would help to make biology a mature science, and his *Descent of Man* (1871) would have a profound impact on how humans viewed themselves (see chapter 4). The full implications of Darwinian reasoning for the concept of race are only now beginning to be understood. Before Darwin, and thus for the vast majority of human history, no scientifically correct understanding of human diversity could have existed.

CHAPTER 1

The Earliest Theories
Race before the Age
of Discovery

Race in Antiquity

Is the problem of racism as old as human relations? If not, then what historical and social developments produced it? These questions have been answered in two different ways. One school of thought suggests that the use of race to make predictions about the individual or social characteristics of people is a relatively modern development, whereas the other suggests that race and racism are fundamental to the way humans think. The former school tends to argue that racism is socially constructed and thus can be socially deconstructed, whereas the latter argues that racism results from biologically ingrained xenophobia. If racist behavior is genetically programmed, then social and cultural programs designed to retard it are likely to be ineffective. Interestingly, both schools marshal much of the same history and data to support their conclusions. Social construction theorists conclude that before the eighteenth century, physical differences among people were rarely referred to as a matter of great importance. There was some tendency to seize upon physical difference as a badge of innate mental or temperamental difference, but there was no universal hierarchy of races in the ancient world.

Biblical Treatments of Race

The three most significant ancient cultures underlying Western civilization are the Hebrew, the Greek, and the Roman. Obviously, the Bible is one of the most, if not the most, significant documents in the formulation of Western culture. For this reason it is interesting to ask whether the Bible makes any specific claims about the origin and significance of race. Before we consider that question, we must recognize that the modern versions of the Bible have gone through many translations and hence may not completely reflect the original writings, which can be dated to at least twenty-five hundred years ago. However, we have no reason to believe that anyone systematically altered Biblical text to remove or add reference to racial bigotry.

One of the Bible stories most relevant to Western views of race is the tale of Noah and his sons (Genesis 6–10; for the following analysis, I will use the Authorized King James Version, which was translated in about 1611). God commands Noah to gather his family into the Ark, for humankind is to be destroyed for its wickedness. Noah has three sons, Shem, Ham, and Japheth, who all take wives before the Flood. The Flood then destroys all living humans except for Noah's family. After the deliverance of Noah and his family, an incident occurs in Noah's tent that leads to his cursing his son Ham, as well as Ham's descendants ("And he said, Cursed be Canaan; a servant of servants shall he be unto his brethren"; Genesis 9:21–25).

Many Christian denominations have claimed that the curse of Ham was revealed in his descendants by their black skin. However, a careful examination of the Bible nowhere reveals God saying that Ham's descendants will be black. In fact, the "families of the sons" of Noah listed in chapter 10 of Genesis are all described as being in the geographical region today considered the Middle East. There is no mention of any populations that could be legitimately considered located in sub-Saharan Africa.[1] The descendants of Ham do not become black or African until between the second and sixth centuries A.D., when the authors of the Babylonian Talmud decide that Noah's curse must have been to turn Ham's descendants black.[2] The phrase "servant of servants" has been used to justify the enslavement of Africans. There is at least one major Protestant denomination that until recently adhered to a variant of the black curse. The Mormons (Church of Jesus Christ of Latter-Day Saints) saw African people as cursed and allowed them to enter only lower-rung positions in the church. The priesthood, for example, was denied to black Mormons, who have always been few.

Not only is there no specific Biblical mention of Ham's descendants being black, but there are even passages that deny the contention that there was any inherent hatred of or belief in the inferiority of Africans in the Bible. For example, Moses was married to a woman described in Numbers 12:1 as an Ethiopian (in Exodus 2:21 Moses marries Zipporah, the daughter of the priest Jethro of Midian). In verses 12:2–15 Aaron and Miriam question Moses for having married an Ethiopian woman, whereupon God punishes Miriam for having doubted Moses. In the ancient Greek, the term "Ethiopian" did not necessarily refer to someone from the country that bears that name today. Instead, the term was used to describe anyone of African origin (of the "Negroid" race).[3]

Throughout the Bible there are many favorable references to people described as being of African origin (Matthew 12:42 and I Kings 10:1–10 recount the visit of the queen of Sheba to Solomon;[4] Jeremiah 38:7–13 tells the story of Jeremiah's rescue by an Ethiopian). The New Testament does not include any apparent racist ideology either. Christ made it clear that he had come to save all humanity (God "hath made of one blood all nations of men for to dwell on all the face of the earth"; Acts 17:26). It seems clear that the cultural views of the Hebrews concerning racial variation have no implicit "racial" prejudice and are therefore consistent with the views of other ancients. Modern authors, such as J. Philippe Rushton, claim that the opposite is true.

Greek and Roman Views of Race

The Greeks generally felt that all non-Greeks were barbarians. However, non-Greeks could shed their barbarian status simply by adopting Greek culture. There are indications that the Greeks did not develop a racist interpretation of physical differences in other peoples they encountered. For example, in one Greek myth, Perseus rescues the Ethiopian princess Andromeda (remember that the Greeks referred to all Africans as Ethiopians). Andromeda's mother, Cassiopeia, was an Ethiopian queen who had claimed her daughter was more beautiful than the sea nymphs (Nereids). Poseidon punished Cassiopeia's kingdom (somewhere in Ethiopia) by plaguing it with a sea monster, and Andromeda was to be sacrificed to placate Poseidon. Perseus fell in love with Andromeda as he passed by, rescued her from the sea monster using the Gorgon Medusa's head, and then married her. There is no discernible negative interpretation of Africans that can be gathered from this story.

The ancient Greeks, unlike the Hebrews, developed many of the branches of modern science, particularly biology and medicine. Thus, many of their early theories of human differences were not based strictly on supernatural intervention. Although Hippocrates (the so-called Father of Medicine, ca. 450–ca. 377 B.C.) felt that the Greeks were superior to the Asiatics, his explanation for that superiority was environmental rather than genetic: he believed that the infertility of the soil made the Greeks self-reliant. He also suggested that the luscious vegetation and abundant crops in more-tropical environments led to softness and lack of war spirit. Thus, he reasoned that Asiatics were feeble, less warlike, and more gentle. Hippocrates had neither an organized theory of heredity nor any conception of natural selection. His argument is akin to that of the eighteenth-century thinker Jean-Baptiste Lamarck, who felt that environmental, or acquired, characteristics could be passed to offspring. This argument also reappears in Darwinian form in the twentieth century.

Aristotle (384–322 B.C., the so-called Father of Biology) thought that racial differences in physical attributes and temperament were caused by climate. In particular, he thought that the mixture of heat and cold was crucial in this regard. The view that heat and cold played a role in shaping the characteristics of biological forms was derived from Greek cosmology, which recognized air, fire, water, and earth as the essential elements. Aristotle's classification of animals in his *Systema Naturae* was also based on this mixture of fundamental elements within each animal, and his views on human races were a natural extension of his general zoological theory. He even thought that changes in the body's elemental composition were responsible for the process of biological aging.

Aristotle's concept of the scale of nature, the great chain of being (*scalae naturae*), would be central to later ideas concerning racial hierarchy. In his *De partibus animalium,* he classified all animals along an ascending ladder of "perfection," according to the mixture of earth, water, fire, and air within the organism. Following the thinking of his mentor, Plato, Aristotle established the position of each organism along the scale of nature as fixed and unchanging. Earthier animals such as earthworms, insects, and other invertebrates occupied the bottom of the scale; wetter organisms (cold-blooded vertebrates, such as frogs) were next; then came fiery animals such as warm-blooded vertebrates; next came humans; and finally airy compositions such as the demigods and gods occupied the highest rungs of the ladder. Aristotle did not explicitly give the human races separate places along the scale of nature, but during the Enlightenment his concept would be revised in the service of racial hierarchy. Aristotle did think that some people were "natural slaves" and others "natural rulers"; but racial variation was not involved in this determination. Again, though, this concept was later co-opted by racial supremacists to justify chattel slavery.

The Roman Empire came in contact with at least three of the racial groups defined by nineteenth-century anthropology. The institution of slavery was a crucial part of the empire for much of its history, and thus we might assume that racist ideology was inculcated in Roman culture. And indeed some Romans did see themselves as superior to other peoples. For example, the architectural historian Vitruvius (active around 46–30 B.C.) attributed the keen intelligence of Romans to the rarity of the atmosphere and the warm climate (consistent with the earlier, Greek worldview). He also thought that the French, Germans, and Britons were mentally slow. He reasoned that the humidity and cold produced in them *"a sluggish intelligence* [italics in original]."[5]

Of course, modern racial theorists (for example, Rushton) now argue that it was precisely this "chill" that was responsible for the evolution of superior European and Asiatic intelligence. These modern theorists thus have difficulty explaining the Roman claim of low intelligence for northern Europeans. From a Roman perspective, the meager cultural achievement exhibited by these groups at the time was proof of their inferior intellect.

Vitruvius was not the only Roman scholar concerned with the nature of human biology. A racist theory of human diversity was developed later in the Roman Empire by Julian the Apostate, who succeeded Emperor Constantine in the fourth century A.D. In spite of the fact that Constantine had adopted Christianity, including its doctrine of the unity of man, Julian pointed out how different the bodies of the Germans and Scythians were from those of the Libyans and Ethiopians. His observation was probably valid in that human body types do indeed vary from north to south. However, he further reasoned that these people must also differ in their dispositions and intelligence. He classified populations into psychological groups. The Celts and Germans were fierce, while the Hellenes (Greeks) and Romans were humane and inclined to political life. However, at the same time the Romans were unyielding and warlike. Julian found the Egyptians more intelligent than the northern peoples and given to crafts, although by his reasoning the Egyptians could also have been warlike and cruel. Egypt had also once been a great empire, of course. The Syrians were unwarlike, effeminate, intelligent, hot-tempered, vain, and quick to learn.

On the basis of these psychological variations, he argued that these different peoples could not have been produced from one pair of humans (Adam and Eve), since the world had not existed long enough for such variations to have arisen. Julian preferred the version of creation given by Plato (ca. 427–348 or 347 B.C.) in his work *Timaeus:* humans were created from drops of Zeus's blood, and the characteristics of different people come from the lesser deities. Ares passed his character on to warlike peoples, Athene gave others both military ability and wisdom, Hermes was the father of those who were intelligent but not bellicose, and so on. Julian the Apostate proposed that heredity determined an individual's character. Thus the nobles were born to be noble, and the lesser humans were born to be inferior. This syllogism is revealed as racist when we realize that Julian categorized groups of people solely on the basis of phenotype, that is, the visible physical characteristics that result from the interaction of genes and the environment. He believed that these visible characteristics were related to other innately determined traits. It is interesting to note that Julian's views on which tribes were virtuous and which were not had much to do with who was in opposition to Rome at the time. Julian thought that Africans were civilized, intelligent, and mild-mannered, whereas he saw Aryans and Anglo-Saxons as warlike, barbaric, and cruel. Precisely the opposite view emerged during the Age of Discovery.

There seems to be little evidence for the idea that the ancient Hebrews, Greeks, and Romans supported the universal superiority of any particular race. Therefore, racism was not often the cause of discrimination against minorities in the ancient world. For example, in India, high-caste individuals had lighter skins and narrower noses, but there was no consistent correlation between caste and features.[6] In neither the Greek nor the Roman civilization was there a relationship between slavery and race. Certainly, Africans were not seen as predisposed to slavery. Any and all captured people could be enslaved. In Roman times, the Punic Wars (with Carthage), the Gallic Wars (with France), and other wars yielded an enormous number of slaves and produced a slave population eclipsing any in earlier history. Syria, Galatia, North Africa, and Gaul produced the greatest number of slaves for the vast slave system. Roman slavery encompassed members of many racial categories. In this system, the owners had virtually unrestricted power, and treatment was truly barbaric. Such conditions, combined with the numerical superiority of slaves over free men, inevitablyled to large-scale revolts, such as that fomented by the Thracian slave Spartacus, who in 73 B.C. escaped to Sicily and formed an army of forty thousand.

The advent of Christianity in the Roman Empire did not interrupt this slave system, because the New Testament, like the Old Testament, did not specifically prohibit slavery. However, the Bible makes it clear that servants can be freed and should be treated fairly. On balance, the early Christian theorists, such as St. Augustine, asserted the unity of man.

The Jews as a Race in Medieval Europe

We have seen that in Europe prior to the Age of Discovery the notion of the unity of humankind was widespread, but there was a notable exception, the Jews. European persecution of the Jews before the Age of Discovery cannot be distinguished from modern racial prejudice. Jewish persecution clearly illustrates that the idea of race can be socially constructed. The Jews were a cultural group rather than a biologically distinct population (to say nothing of a race). Judaism began in populations living in a region roughly comprising present-day Israel and continuing eastward to the western bank of the Euphrates River. The original converts to the religion founded by Abraham were drawn from several of the tribes of the region (the Canaanites, Amorites, Hivites, Amalekites, Kenites, Egyptians, and Hittites). In addition, further admixture resulted from the Egyptian and Babylonian captivities, as well as from the dispersion of many Jews throughout Europe and Africa (remember that Moses' wife was an Ethiopian). A consideration of a variety of phenotypic markers (eye and hair color, head shape, and blood groups, for example) demonstrates that Jewish populations did not differ significantly from any of the non-Jewish populations in the regions they inhabited. Thus the source of bigotry against Jews cannot be located in any underlying biological differentiation. In modern times, Jews in Nazi Germany had to be identified culturally, despite the exaggerated claims of Nazi racial scientists (see discussion in chapter 8).

The persecution of Jews in medieval times had much in common with modern racism. Jews were regarded as having inferior physical, mental, and moral characteristics. They were accused of poisoning the water, engaging in sorcery, ritually murdering children, and desecrating the Host and religious images. They supposedly had horns, an unpleasant odor, and goatlike beards. They suffered from unknown blood diseases, and Jewish men were thought to menstruate. Supposedly Jews needed Christian blood to cure the blood diseases, hence the charge that they engaged in the ritual murder of Christians.

The animosity toward the Jews began early in the history of the European Christian church. There is some evidence that Scripture (the Gospel according to Matthew) was written to minimize Rome's (Pilate's) role in the crucifixion and to blame it on the Pharisees (the priests in Jerusalem). Nevertheless, Gregory the Great, pope from A.D. 590 to 604, forbade persecution of the Jews. This prohibition lasted for about five hundred years, and during that time peaceful conversion was stressed.

In Spain many Jews occupied prominent positions as statesmen, physicians, financiers, and scholars. Jewish and Moorish scholars contributed to the beginning of the European Renaissance (from the fourteenth to the seventeenth century). Outside Spain the attitude toward Jews became more harsh. In the eleventh century, Pope Urban II's call at the Council of Clermont for the rescue of the Holy Land from the "infidel" (the Muslims) was also turned on the "infidel" residing in Christendom (the Jews). Crusaders massacred the Jews at Worms (located in modern-day Germany) within six months of this decree. In 1215 the Fourth Lateran Council of the Roman Catholic Church, called by Pope Innocent III, enacted codes calling for restrictions against Jews. These codes ordered all Jews to wear distinctive badges (this practice did not originate with the Nazis). In cities, Jews were forced to live in special areas, called ghettos, and were not permitted freedom of movement. In Spain, the peaceful era ended in the middle of the thirteenth century with the waning of Muslim domination in the Iberian Peninsula. Under the Catholic monarchs, Spanish Jews were forced into the lowly position of other European Jews. The roots of the twentieth-century Holocaust were planted with the behavior of Christians toward Jews in medieval Europe.

Did Racial Theory Evolve?

We have seen that anti-Semitism in feudal Europe anticipated some of modern racism's characteristics, but is the stigmatization of one cultural group with racist tactics proof that racism is innate? Modern racist theorists seem to rely on the persistence of racist ideology and ethnic strife in human history as proof that xenophobia and, hence, racism are genetically determined. They do so with precious little evidence. It seems clear that human beings have always noticed and recorded phenotypic differences. Throughout the ancient world, humans also speculated about the source of these differences. Much of what we know about this speculation is derived from cultures in Europe, the Middle East, and Africa, probably because it was in these regions that at least two distinct populations (those classified as Negroid and Caucasoid) came into extensive contact. A cursory examination of the classical writings on human diversity does not indicate the hegemony of racist ideology. The raw materials were indeed present, but the evolution of the racism that we know today would require social, cultural, and scientific developments originating in the Age of Discovery and its concomitant colonialism.

CHAPTER 2

Colonialism, Slavery, and Race in the New World

The absence of rigidly applied theories of racial hierarchy in the ancient world presents an interesting problem for the contention that racism is innate. The absence of these theories is inconsistent with the view of racism as a "natural" or genetically programmed behavior. An alternative hypothesis is that changes in social institutions during the European Age of Discovery, and the subsequent colonial domination and enslavement of non-European populations, were responsible for the development of racist ideology. These events also allowed the growth of modern nation states, increased the trade between these states, created new markets, and brought Europeans into extensive contact with physically and culturally different populations.

Under these new conditions, there are two possible explanations for the origin of ideas of European racial supremacy, one nonracist and the other racist. The first explanation is that racist ideology developed out of an objective examination of human diversity. If, for example, European scholars had fairly compared the biological and cultural characteristics of Europeans and non-Europeans and found the former superior, racist ideology would have been validated. In other words, if Europeans really did have larger heads and larger brains, and if these features did determine intellectual ability, we could not label a scientist reporting these facts as racist.

The second explanation is that preconceived notions of European superiority led to a nonscientific justification for European social dominance. Such justification could then have been transformed into a biological theory based on biased science intended to validate European preconceptions. In this scenario, European technological advantages might have allowed Europeans to establish physical or military domination over populations lacking such advantages. But the creation of a social scheme in which non-Europeans were to be enslaved thanks to those accidental advantages of technological development could not itself be used to support innate European superiority. In addition, scientists could not draw conclusions from the measurement of head size, particularly if they could not establish causal relationships between head size, intellectual ability, and social status. To utilize such measurements and reasoning in support of European supremacy would be racist.

Thus, objective scientific reasoning did not create the body of knowledge required for racial ideology to flourish. In the case of Western culture, social change drove much of the development of the biological concept of race and racism. Such social change would accelerate with the contacts between Europeans, Africans, and American Indians in the Age of Discovery.

The Roots of the Atlantic Slave Trade

Prior to the fifteenth century, Christians in Europe were more interested in converting the Muslims in Africa to Christianity than in making use of any of the continent's resources. Papal interventions in Africa were directed at finding allies against Islam to help preserve Christianity on the continent. Early Christian missionary work in Africa was driven by the principle of the unity of humankind as part of God's creation as exemplified by the writings of St. Augustine. For example, in thirteenth-century France, Pierre Dubois proposed that intermarriage would be more sensible than Crusades against the Muslims. Well-educated French noblemen and ladies should intermarry with the Muslim nobility to convert its members to Christianity and monogamy. This scheme would pave the way for French domination in the Middle East and the Orient.

From the seventh century on, the raiding of enemy territory and the capturing and selling of slaves were durable features of the warfare between Christians and Muslims. The tenth and eleventh centuries saw the slave market favor the Muslims (for example, in Spain). After the twelfth century the situation was reversed as Christian naval and military pressure against the Muslims of the Mediterranean increased. This imbalance left the Christians with a growing number of slaves to be sold, many of them Africans from south of the Sahara. By the fourteenth and fifteenth centuries, Christians dominated the slave trade. In the fourteenth century, the Genoese were already selling wheat and slaves along Middle Eastern trade routes. Their main competitors were the Venetians and the Portuguese. In 1444 the Portuguese held the first public auction of African slaves in Lagos, and by 1455 the island of Arguin had become a fortified slave-trading center. In 1466, with the discovery of the Cape Verde archipelago, the king of Portugal gave the settlers there a monopoly on the African slave trade, with the intention of creating a labor force for the sugar plantations. In the same period, the Venetians increased their imports of black labor such that by 1470, 83 percent of the slaves in Naples were black Africans from south of the Sahara. There were also African slaves in Sicily.

In accordance with the law of supply and demand, the increase in the supply of African slaves in European markets drastically lowered their price. As low-priced commodities, these slaves were used in the most brutal and demeaning way. They were thought of as particularly well suited for rigorous agricultural labor, perhaps because they had evolved closer to the equator and thus had skin pigmentation better suited for protection from sun damage, and body proportions that allowed for superior heat dissipation. In addition, they were probably physiologically better suited to heat exposure than were Europeans. These characteristics resulted from both genetic and environmental preadaptations. It is important to note that these adaptations have nothing to do with being black or African, per se; rather they are to be expected of any population that evolved in the equatorial tropics. But in this pre-Darwinian period, before evolutionary reasoning and its ability to explain human physiological variation, these characteristics of Africans were used as justification for their enslavement. The theory of natural law tended to see all things as created by God with a purpose in the hierarchy of nature. Was not the African, a beast of burden, delivered by divine Providence to labor for the benefit of the noble and Christian European? The African rapidly came to be seen as an inferior human being. This shift in social thought began in southern Europe but soon circulated to northern Europe as well.

Effects of the Early Slave Trade on Race Theory

One of the first new racial theories to be promulgated was polygenism, which held that the human races had been created separately. Paracelsus, the Swiss physician, alchemist, and chemist, authored one of the first polygenist theories in 1520. He argued that the children of Adam inhabited only a small portion of the earth and that Negroes and other peoples were of an entirely separate origin. His rationale for this argument was the story of Cain and Abel. Cain had been marked by God and must have married a woman from outside the lineage of Adam and Eve. Genesis seems quite clear in describing at least two lines of descent: Adam and Eve to Seth, and Adam and Eve to Cain. Paracelsus describes the Seth lineage as leading to the whites of the world, and the Cain lineage to the nonwhites.

In 1591, the Italian philosopher Giordano Bruno authored a similar theory. He argued that no thinking person could imagine that the Ethiopians had the same ancestry as the Jews. Thus, either God must have created separate Adams, or Africans were descendants of pre-Adamite races. He felt that exploration and commerce were destroying natural barriers and bringing together people originally meant to be separated: there was by this time a large African slave population in Naples (Bruno was born in Nola, near Naples). Another Italian, Lucilio Vanini, argued in 1619 that, because of their color, Ethiopians must have been descended from apes and that they had once walked on all fours. Ironically, Vanini unwittingly spoke the truth: *all* humans had indeed descended from apelike ancestors. This idea would later be proposed by Darwin, who suggested that Africa must have been the original habitat of humans, owing to the presence there of our nearest relatives, the apes.[1] Isaac de La Peyrère, a French Protestant, revisited the pre-Adamite race theory in 1655, arguing that the pre-Adamite races had given rise to the people of Africa, Asia, and the New World.

The perceived inferiority of sub-Saharan Africans was also becoming part of the cultural folklore of Europe. Italian (and emerging European) racism was examined by William Shakespeare in *Othello*, written in 1604. That racism was a topic of this major work indicates that racist ideas concerning Africans must already have spread to England. Elizabethan literature abounded with lecherous and degenerate black men. The presentation of Moors in Elizabethan drama before *Othello* consisted of only foolish or wicked characters. For example, there was Muly Hamet, the "Negro Moore" in George Peele's *Battle of Alcazar* (ca. 1588), probably the earliest Moor villain; Aaron, the "barbarous Moore" in Shakespeare's *Titus Andronicus* (1593–1594); and Eleazar in Marlowe's *Lust's Dominion* (1599). These characters were used to fulfill the dramatic expectations of this period, whereby a man's color revealed his villainy.[2]

Some classical analyses of the meaning and importance of *Othello* do not consider racism an important component of the text. Clearly, Shakespeare understood his audience's knowledge of anti-African stereotypes and perhaps used these same stereotypes to challenge their validity in his writing. Outside the theater, the negative stereotypes remained, particularly in regard to African sexuality. Sir Francis Bacon referred to "an holy Hermit . . . that desired to see the Spirit of Fornication, and there appeared to him, a little foule ugly Aethiope."[3] Classical criticism saw *Othello* as a story of opposites in human nature but spent little time on the nature of racist ideology. *Othello* was published fifteen years before the first African slaves appeared in Jamestown in the Virginia colony. Chattel slavery in the Americas would be instrumental in the further development of theories of racial hierarchy, as we shall see.

The Spanish Conquest of the Indies

The trans-Atlantic slave trade was made possible by the Spanish conquest of the West Indies and by the subsequent colonization of the New World by the European powers. Westward exploration became necessary because of the stability and power of the Ottoman Empire to the east. The strength of the empire forced Spain and Portugal to find new trade routes to reach the East, and thus the Spanish court was receptive when Christopher Columbus presented his theory that the Indies could be reached by sailing west. Instead, of course, Columbus found the islands of the Caribbean and the Arawak tribes living there in communal social groups. These groups did not view material wealth in the same way the Europeans did. The Spaniards' experience with these populations would greatly stimulate the growth of racist thought in the service of white supremacy.

Having failed to reach the East Indies, Columbus wished to justify his voyage by bringing back some commodities of value from the new territories (particularly gold). This ambition was thwarted by the fact that the islands did not contain much readily obtainable gold or other valuable mineral resources. Having failed to secure gold, spices, or any other form of wealth, he subjected many Indians to slavery. By 1495, fifteen hundred Arawak men, women, and children had been enslaved. Columbus and his crew selected the five hundred best of these for transport to Spain, and two hundred died en route. In the new province of Cocao, on Haiti, Columbus ordered everyone older than fourteen to collect a certain quantity of gold every three months (an impossible task because there was very little gold). Those who managed this feat were given copper tokens. Indians found without a copper token had their hands cut off, and they bled to death. Soon after, Arawaks began to commit mass suicides with cassava poison. Parents killed their infants to save them from the Spaniards. In two years, half of the population of Hispaniola died. When it was clear that there was no gold on Hispaniola, Indians were taken as slaves onto plantations. By 1515, it is estimated that only fifty thousand Indians remained; by 1550, that figure had dropped to only five hundred! One account in the year 1650 reports none of the original Arawaks or their descendants on the island.[4]

Initially, the Catholic kings of Europe discouraged the mistreatment of the Indians, but soon they succumbed to the allure of wealth in the New World. The stage was set for the implementation of human slavery on a scale that the world had not yet known. In 1455 the pope had given Portugal the right to reduce to servitude all "infidel" people. Portugal sought to claim the New World under this papal directive. In 1493 a series of papal bulls divided the area that became South America into spheres of influence for the Portuguese and the Spanish (the east to Portugal, the west to Spain). But papal authority was not enough to stifle the territorial aims of the other European powers. The voyage of John Cabot (b. Giovanni Caboto, in Genoa, ca. 1450) for the English to North America in 1497 was followed by similar French expeditions. In 1580 the English government countered papal bulls with the principle of effective occupation as the determinant of sovereignty.

Whereas the Catholic Church was initially interested in whether people of the new territories could be effectively converted to Christianity, secular interests were more concerned with the production of wealth. Driving down the cost of labor in the colonies was the most effective way to achieve the goals of the European states. As a result of these competing religious and secular interests, the humanity of the Indian peoples was debated theologically and scientifically. The initial discussion concerned whether the Indians in the New World were humans or rather beasts intermediate between humans and animals. Beasts could and should be put to profitable labor, whereas humans might be brought into the fold of Christianity. The agitations of the Spanish missionary and historian Bartolomé de Las Casas were crucial to the church's decision that the Indians were indeed human. Ironically, he offered the colonial powers an alternative solution for protecting the Indian tribes, the importation of slaves from Africa.

The Demographic Impact of the Slave Trade on Africa

There has been much speculation and debate about the demographic impact of the slave trade on Africa. This debate is also relevant to racist theories about the general "inferiority" of Africans. The classical response to concerns that the problems of African Americans resulted from the degraded conditions of slavery was to compare their lot to that of their "free" brethren in Africa. Modern IQ theorists routinely compare Africans to African Americans to try to prove a generalized deficit in African intellect. In the scheme of these theorists, the degraded conditions of blacks in the United States match the lack of social progress of blacks in Africa.

However, if we properly understand that the African slave trade affected both those who were enslaved and those who remained, we can see that this comparison is not relevant. The reduction of African population due to slavery opened the way for the colonization of Africa. Consider just how many lives were lost to the slave trade. There are three estimates of the magnitude of the trans-Atlantic slave trade in Africa from 1450 to 1870, ranging from about 11 million to 30–60 million to as high as 100 million people enslaved. In addition, the sub-Saharan slave trade included at least 9 million people who were part of the Muslim trade to the east. The loss of individuals of reproductive age slowed growth on the African continent. One estimate takes the 19 million persons (10 million to North America and 9 million to the Muslim trade) removed from Africa and assumes a per capita increase in population half that exhibited by the North American slaves to postulate that the population of Africa at the end of the period of the slave trade was 99,420,000 less than it would have been had the slave trade not existed.

This realization also affects our understanding of the origin of European economic supremacy, particularly when we examine how the European populations benefited from the slave trade. In 1600, Africa and Europe had populations estimated at 55 million and 100 million, respectively. Europe's and Asia's populations doubled between 1600 and 1850, whereas the African population grew by only 30 percent. The loss of potential population also contributed to the stagnation of African economic growth and made the continent vulnerable to colonization by European powers (which further accelerated the stagnation). The fact is that African labor fueled the economic growth of Europe and its North American colonies, economic growth that in turn allowed the European populations to increase in number and in fitness at the expense of sub-Saharan Africans. This is the rationale for the charge of genocide in conjunction with colonialism and the slave trade. Africanist scholars such as Walter Rodney and Samir Amin have named the slave trade as the root of the present-day problems faced by African nations.[5]

Slavery Provides New Labor for the Americas

American Indian resistance to enslavement and the subsequent decimation of the Indian populations meant that the colonial powers had to find another source of labor to develop the New World. The long-standing African slave trade was the only logical possibility to achieve this end. The trans-Atlantic slave trade would have profound consequences for the social and cultural development of the modern world. This trade not only caused the economic underdevelopment and stagnation of Africa but also contributed to the underdevelopment of the descendants of the slaves throughout the African diaspora. These two consequences have been primary contributors to the development of modern racial ideology.

The slave trade brought together, for the first time in history, representatives of geographically disparate world populations (Europeans, American Indians [originally Northeast Asians], and Africans). When these groups initially came together in America there were no intermediate populations, that is, populations resulting from interbreeding between the groups. The lack of intermediate populations is important because without them one loses perspective on the continuous and discordant variation in human phenotypic features. For example, when we examine the skin color of people from England and from Nigeria, we observe that for the most part English complexions are less melanic than those of Nigerians. The absence of melanin in the English allows us to see more of their skin capillary blood flow and gives their complexion a pinkish hue. The Nigerians have more melanin, and thus their skin looks brown to black. However, the addition of other populations to this comparison would reveal a continuum from melanic to pinkish complexions. Skin colors do not exist as discrete categories.

The slave trade not only brought together populations that previously had been geographically separated but also brought them together under conditions of manifest social inequality. That is, phenotypic characteristics were used to symbolize social status. A white person could be a lord, plantation owner, merchant, farmer, or even a bondsman. Conversely, almost all blacks were slaves, and almost none could be lords, plantation owners, and so on. The absence of any well-validated theories of heredity meant that no one really understood which features of human beings were innate or which were environmental. Neither was there much belief that the observed physical or cultural features of human beings were mutable. Thus, the concept of discrete racial categories, and the fixed nature of their relations, originated from conditions most extremely illustrated on American soil. The United States became the case that stimulated the interest of philosophers, theologians, and naturalists. How could the unity of humankind be defended, when all common sense suggested otherwise?

Racial Admixture Due to Slavery

The mass implementation of slavery in the Western Hemisphere immediately began to erode the discreteness of racial categories. American chattel slavery in particular had a profound impact on the biological (genetic) makeup of the people we call African Americans. The social history of a population always influences the impact of natural selection, genetic drift, and environment on genes that determine the physical makeup of that population. Understanding how these factors combine to produce the physical characteristics of a population is not always a straightforward procedure. Extreme social hardship, however, is the best example of how social conditions can determine genetic composition. The observations above are inconsistent with the belief that race at the genetic level explains American social stratification.

Those who simplify the American population into black and white miss the fact that no such simple categories can be constructed genetically or culturally within this population. The term "African American" is itself indicative of the process responsible for the creation of this group. African Americans are the product of both biological and cultural admixture (as are ultimately all ethnic groups). Most of the genetic admixture between African Americans and Euro-Americans occurred during the time of chattel slavery. Numerous historical narratives support the fact that many African American women were raped by their Euro-American masters. Even in cases that cannot be easily characterized as rape, the existence of manifest differences between the social statuses of master and slave makes all discussions of consent meaningless. Female slaves who engaged in sexual relations with the master might, for example, have received better treatment or been promised freedom for the children of the liaison (as was reputed to have occurred in the Thomas Jefferson and Sally Hemings affair).

In the United States, individuals of mixed racial ancestry were and still are classified as black by means of what is known as the "rule of hypodescent," whereby one "drop" of black blood made one black. It should be made clear, however, that there is no biological rationale for this rule. *It was a social convention originated to keep the progeny of master-slave liaisons as slaves.*

All human beings receive half of their genetic information from each of their parents (the principle of biparental inheritance). Thus, if one parent were completely African and the other completely European, then half of the child's genes would, on average, be of African origin and the other half would be European. If one knew the origin of all members of a family, one could calculate the average quantity of genes of a given ancestry in any generation.

We can begin to quantify the magnitude of the admixture by comparing the genomes of African and European populations. (The genome is considered to be the entire genetic complement of an individual, the entire genetic message contained in an individual's DNA.) Although European and African populations share the majority of their genomes in common, there are consistent differences between local geographic populations at some loci, that is, at particular locations along a given DNA chain responsible for spelling out specific subsets of the entire genetic message (for example, the locus for eye color). The distribution of alleles (alternative forms of a message found at a particular locus, for example, brown, blue, or green eye color) from such loci in populations can be used to estimate the rate and direction of gene flow between populations.

Allele frequencies in African Americans and Euro-Americans are considered the classic case of unidirectional admixture (one-way gene flow) in modern population genetics. Unfortunately many population geneticists do not realize that the description of this process as "one-way" results from the socially constructed laws of hypo-descent. Clearly some genes of uniquely African origin also must have ended up in the Euro-American population. It is generally felt that between 20 and 30 percent of African American genes originated in the European and American Indian populations (see appendix A). A more detailed discussion of this topic appears in chapters 10 and 11 in conjunction with the concept of the purity of races.

Despite the laws defining the race of "mixed" individuals, some former slaves were able to blend into Euro-American society. An individual with a high percentage of Euro-American genetic admixture could easily appear Caucasian. This was precisely the case with one of Sally Hemings's daughters, Harriet. We know, for example, that Sally Hemings was three-quarters European in genetic composition, and therefore her daughter Harriet (who was fathered by a Euro-American man, most likely Thomas Jefferson) would have been seven-eighths of European genetic origin. Harriet ran away to Washington, D.C., and there married a man of European descent. Her brother Madison, in his memoirs, states that no one ever suspected that Harriet's children had the "taint" of African blood. Thus, the blanket assumption that all people of African genetic heritage were classified as slaves is not quite correct.

The Impact of Natural Selection on the Genetic Composition of the Slaves

There has been much speculation concerning how natural selection might have molded the characteristics of African Americans. Much of this speculation has occurred in the literature and popular myths concerning athletic performance. One much publicized racist comment was made by former CBS sports broadcaster Jimmy "The Greek" Snyder, who stated that the superior athletic ability of African Americans resulted from the fact that "blacks had been bred like race horses." Although Snyder was fired soon after that broadcast, many people felt that Snyder was correct and that African American success in certain sports was directly linked to the biological results of slavery. In *Taboo: Why Black Athletes Dominate Sports and Why We Are Afraid to Talk about It*, Jon Entine concludes that genetic differences resulting from adaptation to local conditions might explain the dominance of "Blacks" in certain sports. However, he consistently confuses the concepts of local populations and socially constructed races, thus not completely escaping the race paradigm.

The argument for African American superiority in sport proceeds as follows. The high mortality that must have resulted during the capture, warehousing, and transport of slaves must have selected stronger and healthier individuals. In addition, "seasoning" occurred once slaves arrived in the Americas, which might also have resulted in selection for disease resistance and physiological stamina. For example, there is strong evidence suggesting that innate immunity might have accounted for differential survival of Africans during the yellow fever epidemics that plagued much of the Caribbean. This example, however, is different from the general argument that suggests that slavery imposed a harsh regime that only the strongest and healthiest individuals survived. Enhanced physiological performance was consistently passed on to their offspring, hence accounting for the supposed superior athleticism observed in African Americans today.

One other portion of the argument often goes unstated and unexamined: that is, the idea that the selection for disease resistance translates into superior athletic performance. If this were true, we could argue that the greater resistance of northern European populations to the AIDS virus is responsible for their dominance in speed skating, swimming, and downhill skiing.

A corollary of the general argument is that the superior athletic performance of African Americans is negatively genetically correlated with intellectual performance. This is a variant of the hypothesis that reproductive performance is negatively correlated with intelligence, an idea spread by psychometrician J. Philippe Rushton. This corollary is severely flawed in several ways. The first flaw is the difficulty of reconstructing the actual conditions under which natural selection could have operated during slavery. Our ignorance about these conditions stems from our lack of strong historical interest in, or concern for, the actual nature of slavery and the rest of the African American experience. In addition, the argument suffers from the main fallacy of all adaptationist just-so stories: the assumption of genetic correlation between physiological performance and specific components of life history. We do not actually know, nor can we experimentally verify, how physiological performance was related to individual fitness during slavery (or even now). For example, a slave could have left more offspring by being intelligent, loyal, and honest or by being sly and morally corrupt. These behaviors might have little to no genetic cause and might be completely independent of physiological or athletic performance. Finally, this argument also lacks any foundation in the sociology of sport.

Table 2.1. Mortality by Age Category aboard the Slave Ship *Coningin Hester*, 1716

Age Group	Females	Males
40–44	0	0
35–39	0.125	0.220
30–34	0.190	0.287
25–29	0.190	0.384
20–24	0.125	0.285
15–19	0.600	0.285
5–9	0	0
Total[a]	0.158	0.277

SOURCE: These data were found at http://www.whc.neu.edu/simulation/histexmp.html and are part of a Web site titled "The Atlantic Slave Trade: Demographic Simulation" (http://www.whc. neu/simulation/afrintro.html). The simulation described at this site was developed at Northeastern University by Professor Patrick Manning, History and African-American Studies; Alexander Mechnikov; Rafael David; and Professor Bryant York, Computer Science.
[a]This is the total mortality for all age classes from the specified sex. These figures are consistent with the range of mortality reported for the Middle Passage. The narratives of individuals involved with the trade suggest that for a given voyage mortality could exceed two-thirds of the "cargo." Philip Curtin estimates that the average losses at sea for the British slave trade were about 17 percent (P. D. Curtin, *The Atlantic Slave Trade: A Census* [Madison: University of Wisconsin Press, 1969], 275).

Any conclusions about the action of natural selection during slavery must be speculative because the mortality data, both on sources and magnitudes, for all the various aspects of the trade are sketchy. For example, the mortality data from the slave ship *Coningin Hester* (Dutch registry, 1716, from West Africa to Caribbean), fail to show clear patterns.[6] The ship carried a total of 183 female and 371 male slaves ranging in age from five to forty-four. Table 2.1 is a calculation of age-specific mortality rates for these individuals. One thing is clear. The data do not suggest that slave mortality on this voyage was mostly related to physiological performance. If human physiology were the sole factor in survival, we would expect to find the highest mortality rates among the oldest and the youngest age categories, which we know are most vulnerable to physiological stress. In addition, we would predict that female mortality would be greater than male mortality. However, we find instead that young adults experienced the highest mortality rates and that males died at about twice the rate of females.

These data suggest instead that behavior, of both captives and slavers, rather than physiology might account for these patterns. Older captives may have survived through their greater experience and patience. Younger captives may have been given special care or quarters on the voyage (in part because of their future value and because they did not present a serious threat to the slavers). However, the mortality rates for both males and females were highest in the age categories that had the highest economic value. Because the years of prime reproduction coincide with the peak of physiological performance in animals, including humans, these slaves should have been the most capable of physiological resistance to the stress of the voyage and, thus, the most valuable to the slavers. Therefore, the high mortality in these age classes is a mystery. One possible explanation is that we see the highest mortality in the categories that were the most likely to resist their imprisonment. Male slaves were considered the most dangerous. Differential punishment, a condition that again depends on the behavior of both the captives and captors, could explain the twofold difference between male and female mortality rates. But we cannot insist on this explanation either. No simple explanation of survival patterns on this slave voyage can be made. In addition, the characteristics required to survive "seasoning" might have been very different from those required to survive any individual voyage. Consequently, any arguments concerning a specific mode of natural selection in shaping the genetic character of African Americans are likely to be fallacious.

There is some need to clear up the confusion concerning the evolutionary processes responsible for gene frequencies in modern African Americans. Natural selection can be defined as the differential reproductive success of favored genotypes. (The genotype is specified by specific alleles found at particular loci.) Properly understood, success has both a survivorship and a reproductive component. However, natural selection alone is not sufficient to explain how changes in the genetic composition of a given population occur. We must also take into account the problem of accidents of genetic processes.

We now know that sub-Saharan Africans have greater genetic variability than all other human populations combined.[7] Slaves were drawn from all over Africa during the Atlantic trade. Those who came to the Americas were mainly from the West Coast. We have no reliable way today to determine the impact of the various sources of mortality during the slave trade on the victims' gene frequencies. For example, we could argue that overweight individuals might have survived the Middle Passage better because of their greater fat stores or, conversely, that overweight people did not survive the process because of the cardiovascular stress caused by obesity.

Certainly, we could argue that the entire process selected for individuals with great intelligence and mental flexibility. However, no one ever argues that African Americans have greater genetic potential for intelligence due to selection during slavery! No one argues that the rigors of being a slave owner, or the hardships of the westward migration in the United States, selected for greater athletic performance. The reason that these arguments are not advanced is that people hold the mistaken belief that only African Americans excel at certain sports. But sociologists of sport have shown that economic opportunity has a great deal to do with who ends up playing certain sports. In the late nineteenth century, many of the best boxers in the United States were of Irish and Italian extraction. Now these groups are rare in boxing. Even basketball was not always dominated by African American athletes. Jewish Americans were the first to succeed at the sport (it was thought that their "craftiness" made them better suited for the razzle-dazzle of the game).[8] If we wish to explain biological factors important to sports performance, we should look to body forms, not to race. For example, the Watusi people would be more likely to excel at basketball than the Pygmy, Yamamoto, or Aleut people because the former are very tall and the latter are short and stout. Body forms do not map racial categories in any consistent way.

We have seen that the Age of Discovery brought about profound changes in Western society and culture. Primarily on the North American continent, it brought together three physically differentiated populations—Africans, Europeans, and American Indians—under socially unequal conditions. The European domination of the continent was made possible by the colonization of American Indian lands and the exploitation of African labor. This scenario could not help but bring changes in the way the politically dominant group treated the subordinate groups. The conditions were set to create a racist culture in the United States, a culture based on a social construction of race. From its inception, this construction was spurious because the underlying biological variation of *Homo sapiens* does not merit the classification of modern humans into different biological categories. It should be made clear, however, that the social category of race is very real and that it has had profound consequences for the historical development of the United States. The next chapter discusses how laws, customs, popular notions, and scholarship have all been involved in producing our modern culture of race.

CHAPTER 3

Pre-Darwinian Theories of Biology and Race

As we saw in chapter 1, the history of biology can be separated into pre- and post-Darwinian periods. Before the Darwinian theory of evolution, biology was a tangle of disconnected empirical observations. During this period, Western explanations of race revolved around religious ideology, mainly the idea that God had created separate races along a scale of perfection. Scientific ideology was not yet independent of Christian theology, and for this reason Western religion and science tended to be in general agreement concerning the significance and hierarchy of human races.

To properly study the biology of human diversity, we must apply principles from taxonomy (the classification of living things into meaningful groups), evolution, population genetics, and biochemistry. Taxonomy may have been one of the earliest biological sciences. It seems that the human mind is preorganized to carry out the process of classification, and that process can be useful if done correctly. However, the criteria upon which a classification scheme is based will necessarily affect the structure of that scheme. For example, one can easily come up with a variety of classification schemes for common household objects. One can group items on the basis of what they are made of: wood, paper, metal, or plastic. One can further classify items by determining whether they are used for work or recreation, or whether they are powered by electricity or some other energy source. If one chooses different criteria, such as purchase date or price range, one will arrive at a completely different classification scheme.

In the same way, all taxonomy is directly related to one's central theory of how (and why) to classify organisms into groups. Household objects are relatively easy to classify, but prior to the advent of evolutionary theory, creating a conscious system that revealed the relatedness of living things was difficult. As we have seen, Aristotle classified living things according to his perception of their proportions of the Greek elements of earth, water, fire, and air, and he created a scale of nature based on this criterion.

The Western concept of race, although socially constructed, was always rooted in "biological" features; that is, the characteristics used to classify individuals into races were always assumed to have a basis nontrivially rooted in human biology. The scale of nature would be used to classify human races through the nineteenth century. Yet the scientific developments that made taxonomy biologically reasonable originated in the nineteenth century and were perfected in the twentieth century. Not until the development of molecular technology was identification of variation at the level of the genetic code possible. Obviously, all previous classification schemes contained errors resulting from an inability to examine the genetic code directly. Some of these errors were egregious, particularly when applied to the problems of the human races.

Eighteenth-Century Naturalists and Race

In 1684 François Bernier, a French physician, attempted to classify all human races and was probably the first European to try to do so formally. Bernier noticed that skin color alone was not sufficient to group populations into races, finding instead that a number of additional morphological criteria were required. However, Bernier had not developed a biological theory that could accurately classify the races. His views were consistent with the idea that there was only one human species and that the different races were thus varieties within the species form. Plato had created the theory of forms, and eighteenth-century naturalists agreed with him that species resulted from divine thoughts in the mind of God. Each species had been designed specifically for its appointed role in nature. Varieties could be formed only by differences in geography and climate. Ample evidence for this principle came from the study of plants. From ancient times, herbalists had recognized the problem of distinguishing true forms from varieties. There was no reason to believe that variation in plants and animals might be qualitatively different.

The foundation of modern biological classification began with the Swedish naturalist Carolus Linnaeus (1707–1778), who developed the binomial nomenclature system to classify and organize plants and animals. Linnaeus was particularly interested in the difference between varieties and true species. In 1735, shortly after arriving in Holland, Linnaeus published *Systema Naturae,* the first of several publications that presented his new taxonomic arrangement for the animal, plant, and mineral kingdoms.

In the tenth edition of this work (1758), he presents his opinions on the classification of the varieties within the human species, aligning them along the great chain of being. He presents two species within the genus *Homo: Homo sapiens* (man) and *Homo troglodytes* (ape); and he describes *Homo sapiens* as comprising four varieties: *H. sapiens europaeus,*

Table 3.1. Eighteenth-Century Naturalists on the Racial Traits of the Negro

	Country and year[a]	Traits examined	Relative to Europeans	Heritable?	Environment?	Separate species?
François Bernier	France 1684	General	Neutral	?	?	No
Gottfried Wilhelm von Leibniz	Germany 1690	General	Neutral	No	Yes	No
Henry Home, Lord Kames	United Kingdom 1774	Skin color, lips, hair, smell	Inferior	Yes	No	No
Johann Friedrich Blumenbach	Germany 1775	Skin color, lips, hair, smell	No ranking of human races	Yes	Yes	No
Samuel Thomas Sommering	Germany 1784	General	Not always inferior	Yes	No	No
Petrus Camper	Netherlands 1786	Skull angle	Inferior	Yes	No	No
Georges Buffon	France 1789	Skin color, smell, intellect	Inferior	Yes	No	No
Christoper Meiners	Germany 1790	General	Inferior	Yes	No	No

[a]The year refers to the approximate date the work concerning the traits of the Negro was first published.

H. sapiens afer, *H. sapiens asiaticus*, and *H. sapiens americanus*. To Linnaeus, the term "variety" meant a group bearing a highly heterogeneous lot of deviations from the species type, deviations in both heritable and nonheritable conditions. The term "subspecies," which refers to heritable (genetic) differences, would not be used until much later in the history of biology. In modern use "subspecies" refers to a group that exhibits a level of genetic differentiation not currently found in the human species.

It is clear that despite problems of terminology, Linnaeus saw a hierarchy of perfection in the physical and intellectual characters of human varieties, with *H. europaeus* representing the apex and *H. afer* the abyss. He describes *H. europaeus* as active and acute, as a discoverer; and *H. afer* as crafty, lazy, and careless. Linnaeus's treatment of Africans was consistent with the general views and approach of other eighteenth-century naturalists examining race. Table 3.1, which summarizes these views, shows that although naturalists examined a variety of character traits, they generally agreed that the Negro was inferior, that this inferiority was mainly inherited (genetic), that environment played a small role, and that Negroes were members of the same species as whites. Although the naturalists disagreed about the relative position of other races, they concurred that Europeans occupied the highest position on the scale of nature. Some of the views and techniques of these researchers are worth reviewing.

Georges Buffon (1707–1788), a French naturalist, saw the white race as the norm, and thus the attributes of all other varieties of humans needed explanation. For example, the dark color of blacks resulted, he supposed, from the intensity of the sun in tropical climates and had been inherited in blacks through the generations. He thought that the dark skin of Laplanders and Greenlanders was caused by excessive cold but that the complexions of these individuals might change through time as they were exposed to moderate sunlight. Buffon thought that little genius could be found among blacks. He did suggest in his *Natural History* that slavery was wrong, however.

Johann Friedrich Blumenbach (1752–1840) seems to have been one of the few voices of reason concerning the classification of human beings in the eighteenth century. He believed that previous categories of human races had been arbitrary in both number and definition. Blumenbach thought that it was possible to identify five main varieties of humans: Caucasians, Mongolians, Ethiopians, Americans, and Malays. He did point out that these blended into each other in imperceptible ways. He based these varieties on the examination of a number of physical traits, such as skin color and cranial measurements. He felt that differences in these traits had been caused by environment and was careful not to imply in his work that the races or varieties within the human species could be in any way ranked. He particularly took exception to the idea that Negroes were inherently stupid. Blumenbach knew of a Negro mathematical genius (probably "Negro Tom") and kept a library of books authored by Negroes.[1]

The Dutch anatomist Petrus Camper (1722–1789) was the first to attempt to quantitatively analyze human facial features. Like others before him, Camper used skulls for his analysis, but his work differs in that he compared human facial features with those of other animals. Camper set up the skulls in an arbitrary position, such that the orifices of the ears and the lowest part of the nasal aperture were in the same horizontal plane. He then drew a line between the front surface of the first incisor and the forehead, and the angle between that line and the horizontal plane became the facial angle (see figure 3.1).

Camper used this technique to measure the facial angles of a monkey

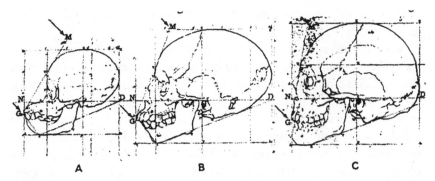

Figure 3.1. The facial angles of Petrus Camper: *A*, a young orangutan; *B*, a young Negro; *C*, a "typical" European. From J. R. Baker, *Race* (Oxford: Oxford University Press, 1974), 29.

(42°), an orangutan (58°), a young Negro (70°), and a European (80°). Camper felt that the facial angle was of taxonomic importance, and he concluded from these figures that the Negro was closer to the ape than the European was. It is not at all clear how many skulls Camper examined to calculate these values. His description of his work suggests that these values were certainly not representative of the groups analyzed (he may have used only one sample). In addition, it is not clear how the values of the facial angle can be said to support the idea that Negroes were closer to apes. His reasoning was attacked even in its own time. For example, Blumenbach mentioned that in his own collection of skulls he had a Lithuanian skull and an Ethiopian skull that showed the same facial angle but were not at all alike in other features.

Jeffersonian Naturalism and the Negro

Of all the Americans of the late eighteenth century who wrote and spoke about race, probably no one had greater influence on the public than Thomas Jefferson (1743–1826). For this reason alone, it is important to report his views (particularly those concerning African Americans). Jefferson, the third president of the United States (1801–1809) and author of the Declaration of Independence, is considered by some as one of the most brilliant men in history. He had a variety of interests: he was a philosopher, educator, naturalist, politician, scientist and inventor, architect, musician, and writer. He was the foremost American spokesman for democracy of his day. Yet it was precisely the conflicts among his ideals of democracy, his practices as a slaveholder, and his philosophy of race that created his greatest intellectual and personal shortcomings.

Jefferson was a creationist, meaning that he believed all living things as they then existed had been created by God in their present form. He felt that all human beings had "unalienable rights" (among these life, liberty, and the pursuit of happiness) and that the Creator had given these rights to all human beings. He also felt that all human beings had been "created equal." But in contrast to this "equal creation" doctrine were Jefferson's observations of his own slaves. He felt that the evidence obtained from his observations, along with those of other scholars and naturalists, could leave no doubt as to the inferiority of Africans. Thus, he reasoned that between the time of the creation of humans and his own time something had occurred to lead to the present conditions and attributes of Africans. Jefferson felt that although Negro inferiority was obvious, the criteria by which it might be established would have to be determined by science. Jefferson did not think that the institution of slavery could account for the degraded condition of Africans. He argued that the ancient slavery of the Greeks to the Romans had still produced Greeks of great accomplishment. Although Jefferson argued against slavery, his argument was in reality more concerned with the impact of slavery on whites than on blacks. He was particularly concerned with its impact on white morals and the laziness and intellectual erosion of the southern aristocracy. These conditions, he felt, would inevitably lead to the bloody overthrow of the slave system, with the resultant destruction of the white race. He thought that the condition of Africans in Africa was further evidence of black inferiority (although he pointed out that the correct comparison was between whites and blacks in the New World). In contrast, he felt that American Indians were naturally equivalent to whites, and he argued against scientific pronouncements against American Indians made by scholars like Buffon.

Jefferson analyzed and compared the characters of various peoples. He thought that blacks had a moral sense equal to that of whites but that the former lacked intellectual ability (consider, for example, his views on African American inventor and mathematician Benjamin Banneker).[2] Even bravery, he thought, was equally distributed between blacks and whites, but he attributed blacks' bravery to "lack of forethought" on their part. He believed that the poetic abilities of Phillis Wheatley were derived from the moral sense, not from the intellect. Jefferson also made it clear that he felt Africans could not inhabit the United States with Europeans and American Indians as a free people. He felt that if African slaves were freed, their resentment of their former owners and their natural inferiority would make it impossible for them to blend into American society. In addition, he suggested that the European's aversion to Africans would make African integration impossible. To Jefferson, this aversion was at first physical—he commented on the offensive color and disagreeable odor of Africans. He also claimed that Africans had bestial sexual natures. This claim is most incredible in the light of the historical, and now biological, evidence that Jefferson took his slave Sally Hemings as a concubine and fathered at least one and probably five of her six children.[3]

Although Jefferson concluded that blacks were generally inferior to whites and Indians, he was never clear on whether this inferiority might relegate them to separate species status. Species status would, of course, have been inconsistent with his doctrine of the equality of man. Jefferson never succeeded in resolving this contradiction, but he did not go as far as the French philosopher Voltaire, who claimed that the races occupied separate positions on the scale of nature.

Naturalists Debate Race

Although eighteenth-century naturalists insisted that there was a hierarchy of human races, they saw all races as members of the human species. However, the beginning of the nineteenth century saw a resurgence of polygenist thinking, particularly in the United States. Table 3.2 shows the views of important nineteenth-century naturalists concerning the Negro race. These views are quite different from those of eighteenth-century theorists. There is no ambiguity here concerning the supposed inferiority of Negroes, the supposition that this inferiority is innate, and the assumption that the environment plays no role. There was, however, some debate concerning whether Negroes represented a distinct species from the other human races. Again the specific efforts of the polygenists to demonstrate and popularize these assertions are worth examination.

The issue of the unity of human races was also featured in the great Paris Academy of Sciences dispute of 1830 between Baron Georges Cuvier and Étienne Geoffroy Saint-Hilaire. The real subject of this discussion was whether all animals had a unity of form (Saint-Hilaire) or whether discrete and fixed animal plans existed (Cuvier). Thus, Saint-Hilaire was arguing a protoevolutionary theory, whereas Cuvier was defending creationist views of the origin and fixity of life. Cuvier used the example of human races to support his idea of the separateness of body plans and the immutability of species. He also made it clear that he saw Africans as an inferior variety of the human species.

When Saartjie Baartman (called the Hottentot Venus) died in Paris, Cuvier compared her physical attributes to those of an ape. He concluded that

Table 3.2. Nineteenth-Century Naturalists on the Racial Traits of the Negro

	Country and year[a]	Traits examined	Relative to Europeans	Heritable?	Environment?	Separate species?
Charles White	United Kingdom 1799	Skulls, sex organs, sexuality	Skulls smaller, sex organs larger	Yes	No	Yes
Samuel Stanhope Smith	United States 1810	Skin color, general	Inferior	No	Yes	No
James Prichard	United Kingdom 1813	Skin color, civilization	Inferior	Yes	Yes	No
Sir William Lawrence	United Kingdom 1823	General, civilization	Inferior	Yes	No	Yes
Georges Cuvier	France 1831	General	Inferior	Yes	No	Yes
Samuel Morton	United States 1849	Skull volumes	Inferior	Yes	No	Yes
Louis Agassiz	United States 1850	Skin color, smell, intellect	Inferior	Yes	No	Yes
John Bachman	United States 1855	Fertility of hybrids	Equal	Yes	No	No
Josiah Nott	United States 1857	General	Inferior	Yes	No	Yes
George Gliddon	United Kingdom 1857	General	Inferior	Yes	No	Yes
Paul Broca	France 1862	Skeletal features	Inferior	Yes	No	Yes

[a]The year refers to the approximate date the work concerning the traits of the Negro was first published.

because of her thick lips she had an apish character. Her skeleton, pre-served at a museum in Paris, was later measured for anthropometric char-acters by French surgeon and anthropologist Paul Broca in 1862. Not surprisingly, the measurement of her features discredited many false criteria in use at that time to "objectively" rank human races. For example, Broca attempted to rank human races by analyzing the ratio of the length of the lower arm bone (the radius) to the length of the upper arm bone (the humerous). He assumed that the more apelike races would show a higher ratio. However, he found blacks at 0.794 and whites at 0.739; but Eskimos were at 0.703, Australian Aborigines were at 0.709, and the Hottentot Venus was at 0.703! Thus Broca's attempt to use this anthropometric measurement to establish an unambiguous hierarchy of human races failed.

The Four Horsemen of Polygeny

The polygenist movement in the United States relied on the work of four men: Louis Agassiz (the theorist), Samuel Morton (the experimentalist), Josiah Nott, and George Robin Gliddon (the last two were popularizers). Louis Agassiz (1807–1873) was one of the best informed and most capable biologists of his day. Agassiz believed in epochs of creation, over which organisms tended to become more complex and better suited to their environment through a series of independent acts of creation by a Supreme Being. This theory supported Agassiz's view that the Creator had also made several different zones of flora and fauna. The boundaries of these corresponded to the relative locations of the human races prior to the Age of Discovery.

Agassiz clearly felt a strong psychological repulsion to Negroes. This was undoubtedly motivation for his polygenist leanings. Despite this dislike for blacks, Agassiz was not a supporter of slavery. He suggested that his polygenist views should not be taken as justification for the mistreatment of any human race (others before him, such as Sir William Lawrence, a physician at the Royal College of Surgeons in London, and Dr. Charles White, also a physician, had found such mistreatment justifiable). However, his views of the abilities of these separate creations were clear. Whites should be trained to use the higher faculties, whereas blacks were suited only for menial labor. He was most repulsed by the prospect of the interbreeding between the races, feeling that this must naturally degrade the moral character of the higher race. This repulsion also led him to reject the definition of species by means of the criterion of interfertility: the ability of individuals within populations to mate and produce viable (fertile) offspring. In contrast, the southern slaveholder and naturalist John Bachman investigated the interfertility of "mulattos" (the racist term for the offspring of an African and a European) and concluded that blacks and whites were interfertile; Darwin would later refer to Bachman's data in *The Descent of Man* as proof of the unity of the human species.

Samuel Morton, a Philadelphia physician, became the leader of the polygenist school and provided purportedly objective data in support of the tenets of polygeny. Morton measured the cranial volume of different races by filling skulls with birdseed or lead shot and then pouring these materials into a volumetric cylinder. In his 1849 tabulation, Morton found that the cranial volumes of the races could be ranked hierarchically, with Europeans on the top and Africans near the bottom. Stephen Jay Gould's classic 1977 reexamination of Morton's conclusions stands as one of the most important revelations of the fallacy of objectivity in science.[4] An examination of both Morton's tabulation and Gould's recalculation utilizing today's knowledge of anthropology and genetics reveals that Morton's results are meaningless: they tell us little about the nature of variation in human cranial capacity or about the genetic sources controlling it.

To demonstrate this, Gould reports a series of unconscious yet systematic errors in the way Morton collected the skulls and analyzed their cranial volumes. Gould remeasured the same skulls utilized by Morton (following Morton's well-recorded procedures). Tables 3.3 and 3.4 compare Morton's results with those of Gould. (Gould did not report his sample sizes, but because he included some skulls that Morton had conveniently excluded from his analysis, we can assume that Gould's sample sizes were probably larger than those used in the original study.) The errors Gould detected in Morton's methodologies fell into the following categories: inconsistencies and shifting criteria that favored expected conclusions, subjectivity directed toward prior prejudice, logical errors in procedure (such as failure to consider alternative causation or correlation), and finally miscalculations and omissions that favored his subjective bias.

The most obvious error (pointed out by Gould) is that Morton did not separate the samples by sex or stature (distinguishing the sex of a particular skull was not always possible). This error is crucial because body size and head size are correlated. Women have smaller heads because they have smaller bodies. The ratio of head size to body size is about equal in women and men. There is no reason to believe that the sex ratios of the populations Morton examined were identical, and there is reason to suspect that the Indian sample contained more women. It was common at that time for governmental and private agencies to pay bounties for Indian scalps.[5] Women and children were often the easiest victims to obtain for this purpose. The ease with which Indian skulls were obtained is also suggested by the number of skulls in each population examined: 30 from the Teutonic family versus 161 from the "barbarous tribes," that is, North American Indians.

Morton also failed to properly consider the fact that populations living in different climates and conditions will show different bodily proportions. Morton recognized this when he grouped different Indian tribes by

Table 3.3. Morton's 1849 Tabulation of Skull Size, by Race

	Volume (cubic inches)	Sample size
Modern Caucasians	92	30
Mongolian group	82	6
Malay group	85	23
American group	79	338
Negro group	83	85

SOURCE: Adapted from S. J. Gould, *The Mismeasure of Man*, revised and expanded ed. (New York: Norton, 1996), 87, table 2.3.

Table 3.4. Gould's 1977 Remeasurement of Morton's Skulls, by Race

	Volume (cubic inches)
Modern Caucasians	87
Mongolian group	87
Malay group	85
American group	86
Negro group	83

SOURCE: Adapted from S. J. Gould, *The Mismeasure of Man*, revised and expanded ed. (New York: Norton, 1996), 98, table 2.5.

stature,[6] but he did not draw the conclusion that stature was correlated with head volume. The fact that environment so strongly influences stature vitiates any idea that genetic controls of head size can be determined by examining physical data alone (the effects of environment will be discussed further in chapter 10).

The problem with Morton's use of physical data becomes even greater when one realizes that there is no established causative relationship between head size and human intellectual capacity. Modern psychometricians maintain that there is a relationship, but their evidence is extremely weak (see chapter 10). They argue that the correlation between head size and reputed intelligence that exists between species should also exist within species. The pyschometricist J. Philippe Rushton makes much of Gould's recalculation. He suggests that even with Gould's recalculation, Morton's data still support the thesis of the superiority of head size in Asians and Caucasians. The difference Rushton calculates of 4 cubic inches (65.5 cubic centimeters) is not "trivial" and cannot be ignored. A mid-twentieth-century study reporting data from white and black males of relatively equal social status, but not correlated for body size, gives cranial volumes of 1,517 and 1,467 cubic centimeters, respectively, or about a 50-cubic-centimeter differential. The average of the two values is 1,492 cubic centimeters, thus the value 50 cubic centimeters corresponds to about a 3.4 percent differential in cranial volume.[7] Rushton and his colleagues ask us to believe that this differential is the origin of the fifteen-point difference in IQ test scores between the races.

After Morton's death in 1851, Josiah Nott and George Gliddon would continue the work of popularizing the polygenist theory. Nott was a native South Carolinian physician, and Gliddon the son of a successful English businessman. Both were proslavery, and both opposed purely religious explanations for the human condition. Nott was acclaimed throughout the South for his lectures on "niggerology." Nott and Gliddon took the battle far beyond the bounds of narrow academic debate. American society was polarizing, and direct action to address the issue of slavery was not far off. Nott and Gliddon published *Types of Mankind* in 1854. It contained none of the "objective" tone of their predecessor, Morton. They also published *Indigenous Races of the Earth* in 1857. This work contained a large ethnographic table that grouped the types of humankind according to three different schemes: geographical (after Linnaeus), physiological (after Louis-Antoine Desmoulins and others), and linguistic (after Louis Ferdinand Alfred Maury and others). Nott and Gliddon divided humankind into eight groups: Arctic, Asiatic, European, African, American, Polynesian, Malayan, and Australian. What is most apparent about their table, however, is the degree of overlap and continuous variation in the three systems. This continuity can also be seen in an examination of the morphology of the portraits of human specimens contained within their ethnographic tables. Thus, as Darwin would point out in 1871, the lack of concordance between taxonomic methodologies used to identify races was actually strong evidence for the artificial character of races. Nott, Gliddon, and the other polygenists could not have seen this, however, for their operational paradigm was the separate and distinct creation of racial groups.

Frederick Douglass and the Polygenists

The abolitionist leader Frederick Douglass addressed the claims of the polygenists in "The Claims of the Negro Ethnologically Considered," an address delivered at Western Reserve College on July 12, 1854. In it he examined both the scientific methods and the political motivations of the polygenists (particularly Morton, Nott, Gliddon, and Agassiz). Douglass pointed out that the fundamental pillar of polygenist thinking and of slavery was the idea that the Negro race was not part of the human family. For example, the law in slave states did not distinguish between Negroes and

other property, such as domestic animals or chairs. Douglass clearly articulated the characteristics of humans that are shared by all races and not
exhibited by animals (anticipating many of Darwin's later arguments in
The Descent of Man). Among these, Douglass included the use of hands,
speech, higher emotions, the ability to obtain and retain knowledge, and
adaptability to different environments.

In his address, Douglass also examined the specific claims of Morton in
Crania Americana, which had been published in 1839. He took particular
exception to Morton's claims concerning the racial identity of the ancient
Egyptians. For Morton, none of the accomplishments of ancient Egypt
could be attributed to Negroes, for that would clearly grant intellectual capacities to Africans unaccounted for by polygenist racial theory. Douglass
advanced the idea that in fact Egypt was a multiracial society lacking the
modern skin color prejudice that existed in the United States and Europe.
None of the Egyptologists of his time supported him on this assertion, but
we now know that Douglass was correct.[8]

Douglass went on to explain how the intellectual achievements of different races are interpreted to suit polygenist theories. He noted that although
Europeans, such as the Irish, might raise their status through education, an
intelligent black man is always supposed to have derived his intelligence
from his connection with the white race. Such suppositions clearly contradicted the rule of hypo-descent. If one drop of Negro blood made you a
slave, how could one drop of white blood make you intelligent?

Douglass also presented the real error in the intellectual content of the
polygenist argument: exaggeration of the differences between the Negro
and the European. "If, for instance," he said, "a phrenologist, or a naturalist undertakes to represent portraits, the differences between the two
races—The Negro and European—he will invariably present the highest
type of European, and the lowest type of Negro."[9]

Finally, Douglass described the impossibility of legitimately comparing
the innate abilities of different races in a society that maintained such disparity in the physical conditions in which the races lived. This is precisely
the reasoning I have applied in my modern arguments concerning race and
IQ. Douglass must be given considerable credit for having accomplished
this analysis without recourse to modern genetics or evolutionary theory.

Conclusion: Scientific Fact or Political and Economic Realities?

Historians and philosophers of science often ask, How much do social
and cultural factors influence the formulation of scientific hypotheses?
The question can be reversed: How much do scientific debates influence

the course of social and political events? In our case, we must ask how much the institution of African chattel slavery and the conquest of North America were influenced by the debate of intellectuals and vice versa. We have seen that both theologians and scientists made massive errors of logic when they approached the question of race in the pre-Darwinian period. It is not a great leap of logic to assume that if scholars were confused and inconsistent about the nature of race, then the lay public must have been much more so. As we have seen, the institution of slavery and the expropriation of Indian land clearly were instrumental in creating this country. Thus the privileged position of Europeans and their descendants was inexorably linked to the race issue. This privilege in turn influenced the way scientific paradigms concerning the nature of human biological diversity were formulated and researched. Public policy, therefore, could not help but be driven by views concerning the position of the races.

Consider, for example, public policy on the issue of slavery. Abraham Lincoln, in the Lincoln-Douglas debates of 1858, stated that he felt the physical differences between the white and black races would make it impossible for them to live together on terms of social and political equality. Given this reality, he felt that the superior position should be occupied by the white race. Thus it seems that one cannot argue that either intellectual or moral correctness on the part of the country's leaders concerning the slavery and race issues was responsible for the abolition of slavery. More likely it was the combination of the slaves (African Americans) effectively resisting this institution (with the aid of their free allies), and the fact that the institution itself was becoming an impediment to the economic growth of the United States, that caused its end. History records many methods of slave resistance, including work slowdowns, sabotage, arson, mass flight, and rebellion. Slave masters feared for their lives also; even domestic slaves often poisoned masters. The South was always on guard against slave rebellions, such as those of Nat Turner, Denmark Vesey, and finally John Brown (1859). Slaves burned down Dallas in 1860. Hinton Rowan Helper's economic analysis of slavery in *The Impending Crisis of the South and How to Meet It* (1857) clearly demonstrated that even Southerners were beginning to see the economic backwardness of this system. Helper showed how the slave system had impeded the economy of the South. The North had more manufacturing and capital, far greater value in farmlands, and vastly more railroad mileage. New York State had more real and personal wealth than Virginia, North Carolina, Tennessee, Missouri, Arkansas, Florida, and Texas combined (even when slaves were counted as wealth). The slave system benefited only the large slaveholders, but the mass of poor whites (among whom Helper was raised) suffered from poverty, ignorance, and superstition.

The impending crisis over slavery in the United States would not end the scientific and moral debates concerning the nature of the races. However, these debates would be fundamentally changed both by the coming changes in society and by the science used to examine the nature of humans themselves.

PART 2

Darwin and the Survival of Scientific Racism

Theodosius Dobzhansky, one of the great biologists of the twentieth century, once said, "Nothing in biology makes sense, save in the light of evolution." If one accepts his statement as true, one must conclude that all biological inquiry was misguided before the coming of modern evolutionary theory. Charles Darwin was not the first evolutionist, but he provided the first clear explanation of the agency both of natural selection in adaptation and of common ancestry in evolution. Variation was a key element in Darwin's thinking. Natural selection was concerned not with just any variation but with variation that defined the survival and reproductive potential of individuals in the struggle for existence. The power of Darwin's theory was that it was immediately applicable to a wide variety of natural phenomena.

This adaptability was precisely what made natural selection and evolutionary reasoning indispensable to race theory. After all, biologists and physicians interested in racial variation in humans had always been studying an evolutionary problem. Thus, the race concept was immediately caught up in the Darwinian revolution sweeping through biology. We shall see that there was no shortage of individuals willing to rush in where angels fear to tread. Soon after the publication of *The Origin of Species*, Darwin was winning over some of the best minds in natural science. Many wished to immediately apply the principles of natural selection to the development of human culture and society. If morphological features were subject to selection, then why not the emotions and the intellect?

Interestingly enough, the conclusions of these early evolutionary theorists of human variation did not diverge from those of their creationist forefathers. How could this be? Did evolutionary reasoning not stand in opposition to all religious conceptions of nature? Was scientific inquiry not objective and immune to the social agendas of this period? Or had the creationists arrived at the correct conclusion about the superiority of the Aryan/Nordic race, but by the wrong means? The coming of Darwinism was the prerequisite for the demolition of the race concept, but this demolition did not happen right away. We shall see that the development of a correct theory, in and of itself, is not sufficient for correct conclusions; the theory must be properly applied.

CHAPTER 4

Darwinism Revolutionizes Anthropology

History is filled with coincidences. For example, Abraham Lincoln and Charles Darwin were born on the same day, February 12, 1809. The fates of both would forever be entwined with the issue of the character of human races. Lincoln's historical role in the ending of slavery in the United States is well known. Darwin's role is not as appreciated but is, in its way, also important. Neither Lincoln nor Darwin had any intention of becoming involved in this drama; but given the events of the mid–nineteenth century, it was almost impossible for any intellectual or politician to avoid becoming involved in or influenced by the slavery issue, which hinged on the concept of race.

Lincoln's perspective on the issue of slavery was influenced by many of the scientific questions being posed by nineteenth-century anthropologists, particularly in Great Britain and the United States. Great Britain was the preeminent global empire, lording over populations from widely disparate continents. The United States was not yet a completely unified nation: the Southern slaveholding states were still more economically and politically tied to Great Britain than to the Northern free states. The political debate about slavery and the biological debate about human variation had both reached the boiling point by midcentury. Science was beginning to establish its authority in explaining nature in general and in particular. Nineteenth-century European and American anthropologists were wholly focused on describing and categorizing human racial variation and on understanding the significance of that variation. As part of this endeavor, they developed techniques that reified white superiority. Consequently, scientific pronouncements would become increasingly important in informing social, political, and legal debates.

By the mid–nineteenth century, the "humanness" of the Negro had not yet been established by Western science. The humanity of the Negro was a key issue for both American abolitionists and slaveholders, and Lincoln addressed this issue in a comment on the Kansas-Nebraska Act of 1854: "Equal justice to the South, it is said, requires us to consent to the extending of slavery to new countries. That is to say, inasmuch as you do not object to my taking my hog to Nebraska, therefore I must not object to you taking your slave. Now, I admit that this is perfectly logical if there is no difference between hogs and negroes. But while you thus require me to deny the humanity of the negro, I wish to ask whether you of the South yourselves, have ever been willing to do as much?"[1]

In this comment, Lincoln questions the practice of treating Negroes as property, as chattel. This practice was also addressed by Chief Justice Roger Taney in the Dred Scott decision (rendered March 7, 1857). In his decision, Taney held that slaves did not differ from other forms of property. "They [Negroes] had for more than a century before been regarded as beings of an inferior order, and altogether unfit to associate with the white race, either in social or political relations; and so far inferior, that they had no rights which the white man was bound to respect; and that the negro might justly and lawfully be reduced to slavery." At the time of the decision, the American school of polygeny was at its height, supported by equally racist science in Great Britain. Based strictly on the prevailing science of the period, Justice Taney's decision not to differentiate Negro slaves from other forms of property was sound. To disagree with Taney was to disagree with the science that declared a plurality of living human species.

It was precisely because Lincoln thought Negroes were human that he argued that slavery was incompatible with American democracy: "When the white man governs himself," he said, "that is self-government; but when he governs himself, and also governs another *man* . . . that is despotism. *If the Negro is a man, why then my ancient faith teaches me that 'all men are created equal,' and that there can be no moral right in connection with one man's making a slave of another* [italics added]."[2]

It was this reasoning that required the Southern slaveholders to rebel against Lincoln's presidency. They realized, even before Lincoln himself did, that viewing the Negro as a human was incompatible with the maintenance of chattel slavery. If human, then the Negro would have the unalienable rights enumerated in the Declaration of Independence. Lincoln felt that even if the Negroes were not racially equivalent to whites as human beings, they still deserved the protection of the law. Despite all the ravings of the apologists for the Confederacy, historical and modern, the Civil War revolved around two central questions: Was the Negro a human? and Was slavery constitutional? The slaveholders, utilizing the reasoning of the polygenists, answered no to both questions and declared that Negroes could and should be enslaved, for the benefit of both races. If, on the contrary, one answered yes to these questions, then one had to oppose chattel slavery. Lincoln and Darwin answered yes.

Charles Darwin, Reluctant Revolutionary

Charles Darwin never intended to become a key historical figure in the debate on the nature of human races. This issue was certainly not one of his main biological interests. However, while Lincoln helped to forge a new nation, Darwin forged a theory that would lead to a new view of man's position in nature. He forever changed our view of ourselves in a way that no one before, or since, has done. Darwin's idea that new species arose through the agency of natural selection had immediate and profound implications for human biological diversity. His was the first scientifically valid program to address the origin and maintenance of biological diversity. Thus, understanding Darwin's formulation of evolution by means of natural selection is paramount for understanding the further development of conceptions of race in the nineteenth century.

Charles Robert Darwin (1809–1882) was born in Shrewsbury, Shropshire, England, the fifth child of a wealthy, sophisticated, and liberal English family. His maternal grandfather was the successful china and pottery entrepreneur Josiah Wedgwood. His paternal grandfather was the well-known eighteenth-century physician and savant Erasmus Darwin. Both Erasmus Darwin and Josiah Wedgwood were opponents of the slave trade, and Wedgwood had created an antislavery medal (and the seal of the Society for the Abolition of the Slave Trade) on which was inscribed "Am I Not a Man and a Brother?"

Charles Darwin was a promising student throughout his preparatory school years, and he was originally slated for a medical education. However, after losing interest in medicine, he enrolled at Cambridge University in 1827 to prepare for a life as a clergyman. At Cambridge he was strongly influenced by Adam Sedgwick, a geologist, and John Stevens Henslow, a naturalist. Henslow taught Darwin to be a meticulous and painstaking observer of natural phenomena and a collector of specimens. Upon graduating from Cambridge in 1831, the twenty-two-year-old Darwin was appointed to the position of unpaid naturalist on a scientific expedition around the world aboard the English survey ship HMS *Beagle*.

On that voyage, Darwin made many important observations that would later contribute to his idea of the "transmutation of species." He noted, for example, that certain fossils of supposedly extinct species closely resembled living species in the same geographical area. On the

Galapagos Islands, off the coast of Ecuador, he observed that each island supported its own form of tortoise, mockingbird, and finch. The various forms were closely related but differed in structure and eating habits.

Darwin also observed colonialism and slavery firsthand on his voyage. These observations would support his belief in the unity of the races within the human species; yet he would nevertheless become convinced of the superiority of certain races in the struggle for existence. Young Darwin had the opportunity to observe racial slavery in Brazil. His response to the oppression that slavery imposed would be the source of constant division between him and Captain Robert Fitzroy of the *Beagle*. Fitzroy claimed that slavery was good for the slaves and that he had heard slaves telling their masters that they did not wish to go free. At this, Darwin questioned the worth of slaves' responses in the presence of their masters.

Early on in the voyage of the *Beagle*, Darwin became convinced that all humans had originated from the same primal stock. He thought that all races were varieties within the human species, and he drew evidence for this from a case he heard about in Tierra del Fuego. The surgeon of a whaling ship mentioned to him that when lice spread from Sandwich (Hawaiian) Islanders to Englishmen, the lice died in a few days. From this evidence, Darwin inferred that all varieties of humans had specific parasites associated with them, that local parasitic fauna had adapted to the specific human varieties they encountered. He felt that this revelation would be a blow to the apologists for slavery, who held that blacks were a separate species. This was a point he wished to take up with the "London experts" upon his return.[3] However, this co-adaptation argument neither proves nor disproves that there are varieties of humans or that they should be considered separate species, since one can also observe parasites that are co-adapted to particular species hosts. It seems that it was Darwin's intuition, rather than irrefutable objective evidence, that was convincing him of the unity of the human species.

Although Darwin felt that all races were varieties within one species, he did not think all races were equivalent in the struggle for existence. After observing the extermination of Tasmania's aboriginal population, Darwin surmised that the immigration of the white races would spell the extinction of underdeveloped indigenous populations. Natural selection would favor the Europeans in interactions with the natives because of the former's superior characteristics. Darwin saw the higher level of European civilization as an agent of natural selection. Without realizing it, Darwin was proposing that the characteristics responsible for the development of civilization were heritable and that they differed between races, or varieties, within the human species.

After returning to England in 1836, Darwin began recording his ideas about the ability of species to change form in his *Notebooks on the Transmutation of Species*. His explanation for how organisms evolved was brought into sharp focus in 1838, when he read *An Essay on the Principle of Population* (1798) by the British economist Thomas Robert Malthus. Malthus posed the question of how human populations could remain in balance. He argued that any increase in the availability of food for basic human survival could not match the geometrical rate of population growth. The latter, therefore, had to be checked by natural limitations, such as famine and disease, or by social or political actions, such as war. Stated simply, more offspring were born than could possibly survive. Malthus's argument helped Darwin formulate the principle of natural selection. Natural selection is a sound concept, even though Malthus's idea of arithmetical increases in food supply and geometric increases in population was flawed.

Darwin was able to apply the "struggle for survival" principle to all animals and plants, and by 1838 he had sketched a theory of evolution through natural selection. Darwin's theory was first announced twenty years later, in 1858, in a paper presented jointly with one by Alfred Russel Wallace, a young naturalist who had come independently to the theory of natural selection. Darwin's complete theory was published in 1859, in *On the Origin of Species*. This work, which has been referred to as the book that shook the world, sold out on the first day of publication.

Why Did Darwin Get It Right?

The superiority of Darwin's ideas concerning the problem of biological diversity is illustrated in how they successfully predict results not anticipated by other theories. For example, creationist theories of the origin of biological diversity had never adequately explained the development of new varieties of animals and plants within species. These theories had also failed to account for either the appearance of new species or the extinction of existing species. Creationism gives no real explanation of why the trilobites became extinct, why three-quarters of all known animal species are insects, or why most animal species have adopted a parasitic mode of life.

Using Darwin's theory, evolutionary biologists have made sense of directly observable biological phenomena, such as the appearance of resistance to pesticides and antibiotics, our ability to produce new food crops, and the postponement of the aging of laboratory animals. Researchers in the biological sciences have now validated the general principles governing evolutionary change. In particular, they have shown how all observed aspects of the biology of organisms can be reduced to how they solve the core evolutionary problem of maximizing their reproductive fitness. Thus, evolution allows us to make sense of previously disconnected phenomena within the biological world. It is now the unifying paradigm of the biological sciences, without which meaningful progress cannot occur.

Evolutionary theory was not, however, fully articulated when Darwin first proposed it; in particular he lacked a correct theory of heredity. He nevertheless managed to outline the mechanism of evolution sufficiently enough to provide a theory with predictive power in biology superior to that of all others. Darwin's general conclusions were the following:

1. Variation in biological populations is important, but not all variations have evolutionary significance (the principle of variation).
2. Variations are inherited in offspring; thus, on average, offspring resemble their parents (the principle of heredity).
3. Some of these variations play a role in the survival and reproductive potential of the individuals that possess them.
4. There is a struggle for existence that governs the abundance of any species.
5. Variations that account for better reproductive success will be passed on to the next generation.
6. Such favored variants, because of the differential reproduction of individuals that possess them, will through time determine the majority of traits found in any given species. (Points 3, 4, 5, and 6 comprise the principle of natural selection.)
7. In some circumstances, changes in the agents of natural selection will lead to the formation of new species over time (as in the adaptive radiation of Galapagos finches).

Darwinians Tackle Human Evolution

The theory of evolution by natural selection has revolutionized our view of all questions in biology. The fact of evolution has been established; that species do change through time has been corroborated by observations made in sciences outside of biology, such as physics, chemistry, and geology. There is now an abundant fossil record of previous life forms, including our human ancestors. Because the chief mechanisms responsible for evolutionary change have also been validated, particularly natural selection, all thinking concerning human variability not based on the evolutionary paradigm can be dismissed. This is not to imply, however, that all analyses by evolutionary scientists concerning a given phenomenon must necessarily be correct. Evolutionists differ in ability and motive, just as other scientists do. Darwin himself discussed human racial variation, including variations in the emotions and intellectual capacity of humans, as part of his evaluation of the general significance of evolution. Given the infancy of the science and its incompleteness (remember the science began without a valid theory of heredity), mistakes were inevitable. In fact, many racist ideologues invoked, and still invoke, Darwinism to justify their racism.

Darwin's theory met with mixed reviews upon its publication in 1859. But Darwin was a fellow of the Royal, Geological, and Linnaean societies, and he also had the support of many influential scientists in England, in particular geologist Charles Lyell, an important member of the Royal Society. Soon after its publication, *The Origin* was clearly winning favor in England with segments of the society that were frustrated with clerical rule over English political and intellectual life. Within five years, Darwin had won the support of many of the best naturalists in Britain and Europe. In November 1864, nine British intellectuals founded a Darwinian supper league called the X Club (the "X" symbolized that there would be ten members). Among the founding members were Thomas Henry Huxley, Joseph Hooker, John Tyndall, George Busk, Herbert Spencer, and John Lubbock. Later William Spottiswoode would join. Many of these individuals were also members of the Ethnological Society, a group interested in applying evolutionary theories to the origin and diversity of humans. The English sociologist Herbert Spencer (1820–1903) had already, before the publication of *The Origin,* worked out a theory of natural selection relating to social forces. Spencer's first evolutionary ideas are found in his *Principles of Psychology* (1855) and provide the basis for what would later be known as social Darwinism. In fact, the tautological phrase "survival of the fittest" actually originated with Spencer, not Darwin. Sir Francis Galton, Darwin's first cousin, was also interested in heredity, particularly the inheritance of intelligence (see chapter 6).

The Ethnological Society was clearly distinguished from its reactionary rival, the Anthropological Society of London (ASL), which was founded by James Hunt in 1863. The latter society fostered amateur anthropology, which had become a favorite pursuit of Victorian gentlemen. The ASL members were polygenists who felt that Africans bore greater anatomical similarity to apes than to Europeans. Hunt felt that as long as Britain possessed an empire, an understanding of the practical importance of race distinctions was essential. He and the other members of the ASL felt that it was not possible to apply the civilization and laws of one race to another race "essentially distinct."

This view played a key role in their defense of ASL member Edward John Eyre, governor of Jamaica from 1864 to 1866. Eyre had reacted to an 1865 rebellion of black farmers at Morant Bay with a degree of ruthlessness unusual even in the nineteenth century. He had declared martial law, and his troops had gone on a murderous thirty-day rampage, killing 439 black people, flogging at least 600 others, dashing children's brains, ripping open the bellies of pregnant women, and burning the homes of more than one thousand suspected rebels. The Anthropological Society rallied behind its controversial member and drew some harsh political conclusions from the affair. James Hunt told the 1866 annual meeting: "We anthropologists have looked on, with intrinsic admiration, at the conduct of Governor Eyre as that of a man of whom England ought to be (and some day will be) justly proud. The merest novice in the study of race-characteristics ought to know that we English can only successfully rule either Jamaica, New Zealand, the Cape, China, or India, by such men as Governor Eyre."[4]

The ASL also supported the Confederacy in the American Civil War. After the war ended, the society reprinted American material calling for the return of slavery in the South and declaring slavery as the only condition under which black people would do any productive work. The Ethnological Society, although convinced that Europeans were evolutionarily superior to Africans, did not see this superiority as a sufficient reason for supporting African enslavement or colonization. They tended to take a more paternalistic attitude toward "inferior races," such as that expressed by Francis Galton in his writings concerning his travels in Africa. There is little doubt that Darwin's opposition to the polygenist views of the ASL influenced his decision to write *The Descent of Man*.

Evolution and the Unity of Races

In *The Origin of Species*, Darwin provided the basic theory required to understand human racial diversity. However, in this work, he was concerned for the most part not with the origin of race or diversity within the human species specifically but rather with the problem of variation in biological organisms in general. In particular, he was concerned with the nature of variations that might be important in the formation of new species. Darwin was the first to correctly describe the essential mechanism responsible for biological adaptation, and his description would lay the groundwork for subsequent investigation into the problem of which variations in a given species persist as the result of natural selection (that is, which are adaptive) and which persist by chance. Understanding the role of natural selection in producing adaptation does not by itself give insight into the general superiority or inferiority of a given group of organisms, including people. However, it does allow us to dismiss any theories that claim to place human races within intelligently designed hierarchies (such as the scale of nature). After Darwin, any ranking of human beings by specific criteria would need scientific (that is, testable) justification.

At the time *The Origin* was published, Darwin already realized where its logic led in the discussion concerning the origin of the human species. Victorian nobles knew that they towered over their black servants, but Charles Lyell posed the question: "Go back umpteen generations and would blacks and whites find a common ancestor? Itself the descendent of an ape? The very idea would give a shock to . . . nearly all men. No university would sanction it; even teaching it would ensure the expulsion of a professor already installed."[5] Darwin felt that the American polygenist Louis Agassiz would be incensed by the thought of the possibility that Africans and Europeans were related in the same species; in Darwin's words, Agassiz "will throw a boulder at me, and many others will pelt me."[6]

Despite Darwin's earlier fears, the stage was set for a full exposition of Darwin's views by the end of the 1860s. The work of Alfred R. Wallace and German biologist Ernst Haeckel on human evolution, along with Galton's work in *Hereditary Genius,* had convinced Darwin of the need for a full exposition of human evolution. Wallace had first read a paper on human evolution before the ASL in 1868, proposing that humans had evolved through competition between groups, not between the individuals themselves. The hardiest races with the greatest ingenuity and cooperation had prevailed. This argument is "group selectionist"; that is, it implies that an individual must sacrifice his or her potential fitness gains to benefit the group as a whole. Later evolutionary theory would demonstrate that such a scenario could be favored only if the groups were composed of closely related individuals. To Wallace, it would have been natural to assume that members of racial groups were more closely related to one another than to members of other races. Thus, he felt that this struggle would inevitably lead to the extinction of the undeveloped nations with which Europeans came into contact.

Darwin was favorably impressed by Wallace's conclusion, for he had observed this process on his own voyages. Wallace also seemed to realize that cultural evolution operated differently than biological evolution. Neither Wallace nor Darwin seemed to recognize that the results of the interaction of Europeans with the indigenous races could just as easily have been explained by the fact that cultural differences are not inherited in the same way that visible physical differences are. Darwin also appreciated the significance of Galton's work on the inheritance of intellect. For natural selection to mold intellect, intellect had to be heritable and had to play an important role in determining an individual's reproductive success. Darwin held no doubts that both these points were true. Finally, Darwin knew of Ernst Haeckel's view of the primitive human races as ontological stages in the development of the modern advanced Aryan/Nordic race.

The Descent of Man

Darwin's *Descent of Man, and Selection in Relation to Sex* (1871) was incredibly popular. By the 1870s Darwinians had established their views on the progression of the human species as respectable intellectual discourse in Britain. In the book's introduction, Darwin lays out his aims: first, to consider whether humans, like any other species, are descended from a previously existing form; second, to describe the manner of their development; and third, to establish the value of the differences among the so-called races.

Before Darwin, anthropologists had noted the anatomical similarities between humans and various species of apes. The polygenists argued that these similarities were due to the special creation of races of apes and humans along the scale of nature. According to the polygenists, Negroes were closer to the ape, and Europeans were higher on the scale than both. Conversely, Darwin argued that the similarity of apes and humans resulted from the fact that they shared a common ancestor. In chapter 1 he uses the following evidence to support common descent: apes and humans share homologous structures, similar development (fetal development, for example), and similar rudimentary structures. In chapter 2 he describes how natural selection can account for the evolution of humans from an apelike ancestor (or lower form). Chapters 3 and 4 concern themselves with showing that humans are not as different mentally from apes as creationist theory suggested.

Chapters 5–7 deal specifically with the origin and nature of variation within the human species. Darwin is particularly concerned with dismantling the views of the polygenists. Therefore, his analysis tends to highlight the unity of human characteristics. Darwin begins chapter 7, called "On the Races of Man," by placing his analysis outside of the standard anthropocentrism of his era. He thus makes the point that naturalists must analyze human varieties in the same way they would examine variation in any other species. He has already established the principle that the mechanisms of nature, particularly natural selection, operate in the same fashion on the human animal as on all other animals. This principle contradicts the general tendency in the nineteenth century to treat the biology of humans as unique (unique human biology was seen as the result of divine intention to separate humans from animals). Darwin also states that much thinking about race is preconditioned according to the position of the observer: "In India . . . the newly arrived European cannot at first distinguish the various native races, yet they soon appear to him extremely dissimilar, and the Hindoo cannot at first perceive any difference between the several European nations."[7]

Darwin then proceeds to examine the correspondence of human races to the zoological zones proposed by Louis Agassiz, concluding that there is some evidence for the local adaptation of humans to their various climates. As we have seen, Agassiz believed that human races had been created as separate species occupying different climatic zones along with other fauna and flora. However, Darwin states that the real definition of species continuity is interfertility. He dismisses the weakly constructed arguments of Paul Broca concerning the supposed lack of interfertility of European settlers and Australian natives. Darwin then utilizes in his main argument for the interfertility of the human races the data presented by John Bachman concerning the fertility of mulattoes. Darwin suggests that the high fertility of the mulattoes in the 1854 census was inconsistent with the low fertility or even sterility predicted of true species hybrids, such as the mule. On the basis of this inconsistency, Darwin suggests that there is no valid evidence supporting the premise that the human races are not interfertile. Interfertility was, to Darwin, proof of the biological unity of the human species.

Darwin also assails the falsity of defining human races with inconstant characteristics. This procedure was known to be fallacious for taxonomic purposes outside of the human species. He points out its absurdity by simply comparing the number of supposed human races defined by various researchers (see table 4.1). The problem of counting the number of races was created by the inability of nineteenth-century biologists to properly identify phenotypic traits that had taxonomic significance. In addition, biologists of the time did not grasp how various phenotypic traits were inherited and how that inheritance might be molded by environment. Before Darwin, the prevailing view in natural science was that all biological traits had significance because they were the result of divine plan. For example, in 1850 the English physician Robert Knox published a book entitled *The Races of Men*. Knox elevated the principle of race such that it had explanatory power over all events in civilization. Knox thought, for example, that democracy was a particular creation of the Teutonic race.

Table 4.1. Darwin's Summary of the Status of Racial Classification in 1871

Investigator	No. races identified
Julian-Joseph Virey	2
Honoré Jacquinot	3
Immanuel Kant	4
Johann Friedrich Blumenbach	5
Georges Buffon	6
John Hunter	7
Louis Agassiz	8
Timothy Pickering	11
G. Bory de St. Vincent	15
Louis-Antoine Desmoulins	16
Samuel Morton	22
George Crawfurd	60
Edmund Burke	63

SOURCE: C. Darwin, *The Descent of Man, and Selection in Relation to Sex* (London: J. Murray, 1871; reprint, Princeton, N.J.: Princeton University Press, 1981), 226.

Modern biological science, of course, recognizes that something as complex as social behavior is strongly influenced by both genetic and environmental sources. In *Types of Mankind* (1854) and *Indigenous Races of the Earth* (1857), Nott and Gliddon developed similar taxonomic schemes to classify the races of the human species. These works made much of the physical features of the supposed races, particularly the facial angles analyzed by Dutch anatomist Petrus Camper (see chapter 3). These angles were used to illustrate the separate and distinct character of each race. Knox proposed his own theory of facial profiles (see figure 4.1) In *Indigenous Races of the Earth*, Nott and Gliddon provided a series of portraits to illustrate human variation and used these portraits, along with cranial angles, language types, and other data, to represent what they considered the discrete nature of the variation found among the races of humans.

My view of their data suggests the opposite conclusion. The original artists' drawings were superbly done and immediately gave me the impression of the continuity of human variation. Nott and Gliddon portrayed human beings from around the world. The only named portrait in the collection was that of Georges Cuvier, the renowned racist and so-called Aristotle of his age. All the others were labeled, for example, Hottentot, Mozambican Negro, Chinese, Fuegian, Eskimo, or Syrian. This labeling practice reveals the Eurocentric bias of the authors' approach to the characteristics of the races. For example, the authors could either have presented unidentified portraits for each race or found specific named individuals for all the categories they presented.

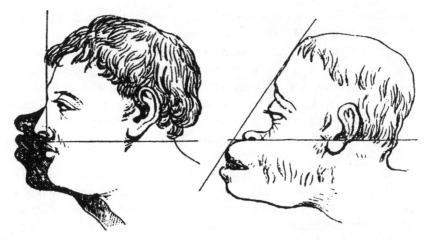

Figure 4.1. The facial profiles from Robert Knox, *The Races of Men.* This figure shows no transition between the sub-90° angle of the orangutan and the 90° angle of the European. The Negro is inferred to have a sub-90° angle. Robert Knox, *The Races of Men: A Fragment* (Philadelphia: Lea & Blanchard, 1869; reprint, Miami, Fla.: Mnemosyne and Co., 1969).

Examining the state of racial classification in the nineteenth century, Darwin concludes the obvious: the heterogeneity in the number of races identified showed that the investigators could not agree on what features of the human being could or should be used to define the category. This disagreement also suggested that the categories must have shown gradation into each other. Indeed, anyone objectively viewing the material that Nott and Gliddon provided in *Indigenous Races of the Earth* could not fail to make such an observation. Detailed analysis of the existing morphological variation within and between these racial categories has forced modern physical anthropology to conclude that no consistent race designations can be constructed, thus validating Darwin's original suggestion (see, for example, the statement of the American Society of Physical Anthropologists on race published in 1996).

Darwin utilizes the previous observation to conclude that the polygenist interpretation of the origin of human races cannot be correct: "Although the existing races of man differ in many respects, as in colour, hair, shape of skull, proportions of body, etc., yet if their whole structure be taken into consideration they are found to resemble each other closely in a multitude of points. Many of these are of so unimportant or of so singular a nature that it is extremely improbable that they should have been independently acquired by aboriginally distinct species or races."[8] Here Darwin states that human populations are more alike than they are different, particularly in a range of traits that would not have been the concern of natural selection. The similarity in these unimportant traits, he says, would not have occurred if human races were separate species. To Darwin, these similarities are the result of common descent. Again, he anticipates a result supported by twentieth-century genetics. We now know that modern humans have not maintained separate enough gene pools to be considered unique evolutionary lineages. This explains the very small amount of overall genetic distance that we currently observe within our species.

Darwin ends his discussion with some speculation concerning the extinction of certain races of humans and the origin of races themselves. He points out that the contact between more "civilized" varieties of humans and "barbarous races" often leads to lower fertility in the latter. He suggests that natural selection acts against the barbarous races through the tools that civilization has given the Europeans. He cites the contact between Europeans and the Sandwich Islanders, showing that between 1832 and 1872 the native population declined by 68 percent. He is, in fact, primarily adducing evidence for the negative biological impact of colonialism on the colonized.

Consider, for example, colonialism's effect on the food supplies of indigenous populations. In Latin America and Africa, colonization overturned the existing system of food production and distribution in the conquered populations. What replaced the indigenous subsistence agriculture was a system designed to grow cash crops for export, which in turn favored the creation of large plantations and the resulting formation of landed oligarchies. Thus, the contradiction of these colonial agricultural economies was that although their agricultural productivity was high, they produced little food for the indigenous populations. In the worst cases, colonies dominated by cash-crop economies became net food importers. The imported food was often too expensive for farm laborers and other poor people, and widespread malnutrition resulted. In addition, as the value of agricultural land increased, the Europeans tended to remove indigenous populations from the best agricultural land, either by direct force or by economic means. The result was long-term stagnation in population growth and, in some cases, the literal extermination of indigenous populations.

Darwin's speculation about the origin of human races revolved around two potential mechanisms: correlated selection and sexual selection. Correlated selection results when selection for a trait also causes differences in an organism's ability on an additional trait or traits. An example of this could be selection for starvation resistance also causing a general increase in survival. Darwin suggested that differences in skin color might not have resulted directly from climate but might instead have been linked to disease resistance. He discussed the case of yellow fever epidemics and the higher survival rates of Negroes in the Caribbean. Again, this observation has been supported by modern anthropological research on the changing demography of the region. Variation in "racial" characters might also have resulted from sexual selection, which refers to a female's choices concerning which males make the best mates. Sexual selection might even conflict with natural selection, as when peahens choose peacocks with large tails. These males could have lower survival in the wild yet higher mating success.

Darwin accepted the principle that the so-called races of humans seemed to differ in intellectual, moral, and social faculties. Again, referring to the general similarities of humans, he stressed the flexibility of these differences: "The same remark holds good with equal or greater force with respect to the numerous points of mental similarity between the most distinct races of man. The Australian Aborigines, Negroes, and Europeans, are as different from each other in mind as any three races that can be named, yet I was incessantly struck, while living with the Fuegians on board the Beagle, with the many little traits of character, shewing how similar their minds were to ours; and so it was with a full-blooded negro with whom I happened once to be intimate."[9] He continued, "The great variability of all the external differences between the races of man, likewise indicates that they cannot be of much importance; for if important, they would long ago have been either fixed or eliminated. In this respect, man resembles those forms which the naturalists label as protean or polymorphic, which have remained extremely variable, owing as it seems to such variations being of an indifferent nature, and to their thus having escaped the action of natural selection."[10]

Darwin made a point of excepting traits that seemed related to reproductive fitness, such as immunity to disease and intellectual capacity. However, he anticipated the results of modern population and molecular genetic investigations of human diversity: that is, that most of the features that are differentiated among the populations of the world have little or nothing to do with reproductive, and thus evolutionary, success.

In addition, Darwin was not concerned, as the ASL was, with constructing or reifying a hierarchy of human races. In his writings he addressed human variation insomuch as it illustrated general principles relating to natural selection and the resultant organic evolution of our species. We know that although he opposed slavery and racism in general, Darwin was not immune to the Eurocentrism of his time. His standard of racial achievement was the Victorian ideal. Much of the material he cites uncritically in *The Descent of Man* is derived from sources that were clearly intent on documenting the superiority of Europeans. Nevertheless, Darwin's scrupulous adherence to the principle of scientific logic and critical thinking makes his general analysis of the problem of variation in the human species vastly superior to the analyses of his predecessors. For example, Darwin questioned Nott and Gliddon's claim, based on their examination of a statue, that Pharaoh Amunoph III had no Negro ancestry. Their claim was consistent with the views of the polygenists, who felt that no one of Negroid ancestry could have held any position in ancient Egypt other than that of slave or peasant. On the contrary, when Darwin examined the same statue, he described the pharaoh as having "a strongly-marked negro type of features."[11] This observation suggests that Darwin was capable of making an objective assessment of the pharaoh's phenotypic traits. Thus, he was not blinded by the "fact" of inherent Negro inferiority.

Throughout his work, Darwin was more circumspect than most of his contemporaries in his application of natural selection to human variation and in his interpretation of the data. His conclusions must be carefully compared to those of the scientists who were ideologically committed to the superiority of the European. We should not have expected Darwin to come to any other conclusion, for Darwin's natural selection was fundamentally at odds with interpretations of nature that held to divine schemes or to hierarchies. To Darwin, human variation was the result of a process of natural selection, which at the ultimate level was unconcerned with socially constructed distinctions. For example, natural selection might maximize reproductive fitness in humans with little regard for anthropocentric distinctions. Thus, although we might value the morality and wisdom of the frugal, sagacious Scotsman who limits his reproductive output, he nevertheless might be acting in opposition to the rule of nature by doing so. And although we might look askance at the "vice and degradation" of the Irishman who reproduces young and has a large family, his behavior favors his reproductive success.

Today, we know that Darwin's intuition about racial variation in humans was essentially correct. The level of genetic variation that exists in the human species is not close to being high enough to allow the definition of subspecies (or races). Darwin did, however, inconsistently use the concept of race in *The Descent of Man*. Although he discussed the taxonomic limitations of the concept of race, he nevertheless referred to advanced races displacing indigenous ones. This inconsistency results in part from his lack of a fully articulated population genetics theory, which could have helped him resolve this conceptual difficulty. Population genetics allows us to apprehend the fact that there are no biological or geographical races within the modern human species. The differences that do exist in human populations result from selection along independent climatic gradients. These gradients include such things as solar radiation, heat, humidity, disease organisms, and so on. In addition, chance events cause variation of gene frequencies, such as those that occur in small, isolated populations. Consider, for example, the high frequency of Alzheimer's disease found in the population of Russians known as the Volga Germans. This population is descended from a small group of Germans who migrated to an area near the Volga River in Russia; these Germans maintained their Germanic heritage and did not intermarry with the native population. Purely by chance, the original migrants had a high frequency of the App-1 gene, now known to be associated with predisposition for Alzheimer's.

Darwin Influences Anthropology

One clear example of Darwin's influence on anthropology in the latter half of the nineteenth century is John P. Jeffries's *Natural History of the Human Races* (1869). Jeffries utilized evolutionary theory to explain the facial angles of the Negro and the European (see figure 4.2). Jeffries proposed that there was a continuous evolutionary progression from the sub-90° profile angle of the Negro to the 90° profile angle of the European. The inference is that the sub-90° angle is a primitive condition and that the 90° angle represents the advanced state. Jeffries's proposition demonstrates that evolutionary mechanisms could indeed be placed in the service of racist ideology. There was no a priori reason to suppose that one angle was original and the other derived. Jeffries's conclusion was based on the prevailing nineteenth-century idea of the inferiority of the Negro.

Such observations suggest that Darwin's views on human variation did not readily win over the anthropological community. In some scholarly quarters, the idea of descent from a "lower" form was accepted, but the unity of modern races was ignored. There was greater acceptance of Darwinism in England than in the United States, which succumbed only slowly. Why did Darwinism not rapidly transform nineteenth-century conceptions of race? This question can be separated into at least two component parts. First, to what degree were Darwin's ideas on natural selection accepted by scientists, naturalists, and the general public in the nineteenth century? Second, to what degree do the views of biologists impact public thinking on race, then and now?

The first question is easier to address than the second. The universal scientific victory of Darwinism really did not occur until the neo-Darwinian

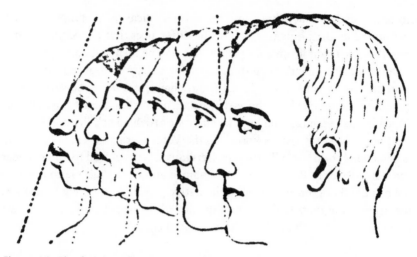

Figure 4.2. The facial profiles from John Jeffries, *The Natural History of the Human Races*. In the wake of *The Origin of Species*, Jeffries proposed that the Negro was an ancestral race and that natural selection had eventually produced the more evolutionarily advanced European. Ernst Haeckel would later advance a similar theory. J. P. Jeffries, *The Natural History of the Human Races* (New York: E. O. Jenkins, 1869), 347. This work can be obtained from University Microfilms, Ann Arbor, Mich.

synthesis, that is, until the unification of Darwin's mechanism of natural selection with Mendelian genetics, a unification that started at the beginning of the twentieth century and was not completed until the 1940s. The acceptance of Darwin's principles varied by country; for example, the grip of Lamarckianism in France held up the acceptance of natural selection until the late 1890s in that country. Lamarckian biology held that although evolution happened, it was strongly influenced by the inheritance of acquired characteristics. This is opposed to the idea that genetic information is inherited and then acted on, pro or con, by natural selection. Soviet Russia was actually the place where neo-Darwinism flourished most in the 1920s. Marxists felt that neo-Darwinism had a strong materialist content and was thus consistent with their own ideology. Stalin's counterrevolution would later derail evolutionary thinking in the Soviet Union by reinstalling Lamarckian thinking in genetics, as illustrated by Lysenkoism. Trofim Lysenko was a Soviet biologist who felt that characteristics acquired from the environment were more important in determining an organism's success than were inheritable or genetic traits.

The second question turns on the degree to which scientists inform public debate about important social issues. The impact of scientific information on ethical, legal, and cultural issues is unpredictable. We know that many uninformed ideas about the biological attributes of race still circulate. At the end of the nineteenth century, both Americans and Europeans were obsessed with the idea of race. The average American citizen lived in a world where the positions of the three major "races" were still manifestly unequal. In the United States, the period from 1850 to 1900 would see the Civil War, Reconstruction, the rise of the Ku Klux Klan and Jim Crow, the wars against the Plains Indians, the war against Spain, a wave of new immigration from Europe, and other events directly related to racial issues. Race, as it was socially constructed, was at the center of American life. It seems naive to think that any amount of scientific scholarship could have altered the racism of this period. In addition, the scientific community was almost wholly derived from European cultural roots. Thus scientists had no direct material motivation for resisting the spread of racist ideology and even less personal motivation. Some did resist, as a matter of intellectual consistency (as in Darwin's case). None, however, played a major public role in opposing racist thinking. The concept of the "white man's burden" was well established. If the scientists could not or would not formulate the question clearly, then it was impossible for them to positively influence the public at large.

CHAPTER 5

Race and Social Darwinism

In 1864 Charles Darwin wrote to his friend the American botanist Asa Gray: "The destruction of slavery would be well worth a dozen years of war."[1] At the conclusion of the American Civil War in 1865, more than 4 million people who had some amount of African blood and were identified as Negroes were freed from bondage. What role would the newly freed slaves play in the reconstruction of the country? Initially the Union occupation force in the South provided some protection for the newly freed slaves, but by 1877 the South would be "redeemed." The new political realities of Jim Crow reinstated the racial theory that had buttressed chattel slavery. Full participation in Southern society was now premised on the power and purity of white blood.

The racist author Thomas Dixon's fictional work *The Leopard's Spots* claimed that freedom would not alter the inherently degraded character of the Negro. The Southern polygenists J. C. Nott and G. R. Gliddon continued to assert that science declared the Negro subhuman, and they were supported by the work of the ASL in England. The Southern states in unison enacted "black" codes, chief among them provisions that defined racial status, forbade interracial marriage, and imposed the death penalty for blacks who raped white women. Virtually all the black codes defined race using the principle of hypo-descent: that is, any detectable African ancestry was sufficient to classify one as a Negro.[2] The lynch rope reinforced political and economic disenfranchisement, whether accomplished "legally" by all-white juries or extralegally by terrorist organizations, such as the Klan.

The solution of the "Negro problem" in the South prepared the United States for western expansion. The conquest of the West would play a major role in the subsequent consolidation of the United States into a world power. To the fashionable Euro-American intellectual of the late nineteenth and early twentieth centuries, the secret of the United States' success resided entirely in the special abilities of the Anglo-Saxon/Nordic race. Mid-nineteenth-century racist theorists had been able to rely on polygenism, but the power of creationist doctrine had been severely compromised by the publication of *The Origin of Species* and by subsequent Darwinian arguments supporting the unity of the human species. A new theory was required, and this would be provided by social Darwinism. Social Darwinism was born in England but spread rapidly to the United States.

Spencerism Comes to the United States

In the midst of the chaos of the Civil War and Reconstruction, events were brewing in England that would have profound effects for American social debate. Herbert Spencer, born April 27, 1820, published a theory of progressive social evolution in 1852, seven years before the appearance of *The Origin of Species*. In two essays he coined the term "survival of the fittest," which would later erroneously be attributed to Darwin. Spencer's major ideas were an eclectic mix of Charles Lyell's uniformitarianism, Jean-Baptiste Lamarck's theory of development, Karl Ernst von Baer's embryology, Samuel Taylor Coleridge's conception of a universal pattern of evolution, anarchism, laissez-faire economics, Malthus's population theories, and the principle of the conservation of chemical energy of James Joule, Julius Mayer, Hermann Helmholtz, and Lord Kelvin. Spencer, a member of the X Club, would co-opt elements of Darwin's views on the role of natural selection after the publication of *The Origin*. Today we know that these incorporations were invalid; yet Herbert Spencer would have far more impact on American social thought than Darwin ever did. In 1864 the *Atlantic Monthly* proclaimed: "Mr. Herbert Spencer is already a power in the world. . . . In America, we now confess our obligations to the writings of Mr. Spencer, for here sooner than elsewhere the mass feel as utility what a few recognize as truth. . . . Mr. Spencer represents the scientific spirit of the age. . . . Mr. Spencer has already established principles which however compelled for a time to compromise with prejudices and vested interests, will become the recognized basis of an improved society."[3]

Virtually every major Euro-American thinker of the latter portion of the nineteenth century was profoundly influenced by Spencer (William James, Josiah Royce, John Dewey, Borden Browne, Paul Harris, George Howison, James McCosh, and the founders of American sociology, Lester Ward, Charles Cooley, Franklin Giddings, Albion Small, and William Graham Sumner). Edward Livingston Youmans wrote to Spencer: "I am an ultra and thoroughgoing American. I believe that there is great work to be done here for civilization. What we want are ideas—large organizing ideas—and I believe there is no other man whose thoughts are as valuable for our needs as yours are."[4] Spencer's influence probably did not extend to African Americans. I can detect no Spencerian influence in W.E.B. Du Bois's excellent essay on the nature of human races in 1897 or in his *Talented Tenth,* published in 1903, although Darwin is clearly mentioned in the former. Spencer, together with Francis Galton, would profoundly influence politicians and industrialists, such as Henry Cabot Lodge, Calvin Coolidge, Teddy Roosevelt, Woodrow Wilson, Winston Churchill, John D. Rockefeller, and Andrew Carnegie.

Charles Darwin himself found Spencer a bore and his ideas repugnant. The fact that the Spencerians dubbed themselves as Darwinians is not reason enough to lay the ugly legacy of Spencerism at the feet of Darwin. Darwin clearly felt that natural selection played an important role in the development of society. This idea is outlined in several sections of *The Descent of Man,* for example, in "The Extinction of Lessor Races" and "Natural Selection on Civilized Nations." In the former section, Darwin was examining real effects of colonialism on less industrialized populations. He did not claim that colonialism was a virtuous activity; rather he saw it as the result of the difference in "mental" abilities, which were related to fitness and, thus, under the power of selection. In the latter section, Darwin discussed the human interference with the actions of natural selection. In the case of medicine, Darwin correctly pointed out that some medical activity counteracts natural selection: for example, the treatment of individuals with otherwise fatal conditions. If individuals who would normally die without intervention reproduce, and their condition is heritable, the frequency of the genetic malady in the human species will increase. However, Darwin made it clear that not intervening to help these individuals went against human moral sense and altruism, both traits that he thought were also the result of natural selection. Utilizing the same reasoning as Darwin, Alfred Russel Wallace concluded that human societies might be selected for cooperation, as opposed to competition.

Spencer's chief American disciple was economist and sociologist William Graham Sumner of Yale University. Sumner achieved his professorship in 1872 and would spread the influence of Spencerism. Sumner was the chief American propagandist of the idea that capitalism was a natural, automatically benevolent, and free competitive system. He wrote that millionaires were the product of natural selection, the bloom of the competitive society. He also felt that human progress was ultimately moral progress, as measured by economic virtues, such as the accumulation of wealth. Sumner deplored any governmental activity that helped the poor, arguing that they were unfit for survival. Other prominent social Darwinians who focused on issues relating to race were Granville Stanley Hall (psychologist and educator), Francis Galton, and David Starr Jordan (president of Stanford University).

Not surprisingly, the American mainstream intellectual resistance to Spencerism in the late nineteenth century was as racist as Spencerism itself. Both sides of this debate attempted to explain the class stratification among whites and assumed the clear inferiority of other races. The main topic of discussion among the opponents of Spencerism was, of course, the status of the Negro. Lester Ward, a professor of sociology at Brown University, divided the races into two categories: those who had historically been favored (the white races) and those who had not (the yellow, brown, red, and black races). He suggested that blacks were more developed than whites in terms of sentiment and less developed, to the same degree, in terms of intelligence. He credited people of the "yellow" race with diligence but believed they were inferior in feeling and intelligence. Ward suggested that it was natural for black men to attempt to rape white women, even under the threat of the lynch rope. He saw this behavior as a biologically ingrained attempt on the part of blacks to "improve" their stock.

Despite this grotesque analysis of race relations in the South, Ward did manage to make some important criticisms of Spencerian ideas. In response to Galton's *Hereditary Genius* (1869), Ward suggested that the supposedly innate superiority of the leaders of nations was the not the cause but the result of their higher social position; that is, that their social position created their apparent superiority. Ward suggested that individuals who grow up in favored environments should be expected to develop both superior physiques and superior intellects as compared to those raised in poverty. Ward's point is important, for it raises the fundamental question of the nature versus nurture debate, which was not often presented in this period. In modern quantitative genetic terms, Ward is addressing three components that contribute to the expression of genetic material: environmental effects, interactions between genes and the environment, and the covariation of genes with environments.

A hypothetical example of an environmental effect is illustrated in figure 5.1. The figure shows the mean survival times of two genotypes within a species exposed to a hypothetical stress. Moving the genotypes from environment 1 to environment 2 alters their mean survival time. Environment 2 is more permissive for survival for both genotypes, and survival times for both improve by 100 percent (the variability also increases proportionately). If we compared these two genotypes statistically, we would find no significant difference between them in either environment.

Consider a real-life example of this type of environmental effect. There are a number of classic cases in which environmental change has impacted

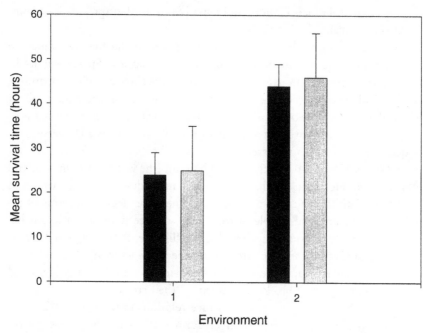

Figure 5.1. A hypothetical environmental effect on phenotype

public health, for example, the eradication of the hookworm *Necator americanus* (called the American killer) early in the twentieth century. This nematode causes severe anemia, particularly in children. In 1902, the American parasitologist Charles Stiles showed that the shiftless behavior, apparent laziness, and low intelligence scores of many southern children, black and white, resulted from severe infestations of this parasite. The cure for these children was simply providing them with flush toilets and shoes. This cure worked for both races; per the example above, it amounted to transferring children from environment 1 to environment 2.

However, the impact of environment on the expression of genes is not always uniform. A hypothetical gene × environment interaction is shown in figure 5.2. In this case, the mean survival time of both genotypes improves in environment 2, but the survival time of the genotype represented by the gray bar improves much more than that of the genotype represented by the black bar. In this case, if we compared the mean survival times of the two genotypes, our result would depend on the environment in which our comparison was made. In environment 1, we would conclude that there was no difference between the gray and black genotypes, but in environment 2 we would consider them statistically different.

The *Necator* case can also serve as an example of this interaction. Dur-

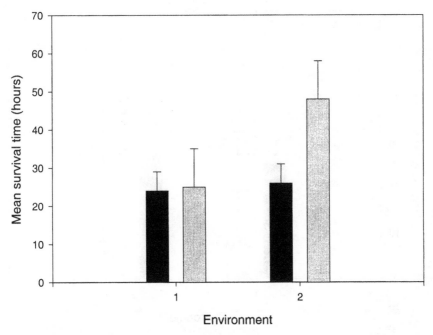

Figure 5.2. A hypothetical gene × environment effect

ing World War I, the National Academy of Sciences performed for the army a set of intelligence measures on whites and blacks, in groups infected by *Necator* and in uninfected groups. The infected blacks showed a 14.7 percent decrease in mean performance, compared to a 20.3 percent decrease in that of whites.[5] If we assume that the tests were measuring a physiological impact of hookworm infection on some component of intellectual function, one hypothesis is that the blacks might have retained some resistance to *Necator*. The genus *Necator* is not native to North America but was brought to North America with the importation of African slaves. Blacks' resistance to the parasite might have been genetic, arising from ancestral selection in western Africa. This hypothesis would be consistent with the observations of Africans' greater resistance to tropical disease compared to that of Europeans.

In the example of the covariance of genes and environments, the third component of gene expression, environment 2 is always benign for our hypothetical organisms and environment 1 is always detrimental (see figure 5.3). Clearly, the mean survival time calculated for each would depend on the distribution of the genotypes between the two environments. If the differential distribution of the genotypes between the benign and detrimental environments is not taken into account, an observed phenotypic difference

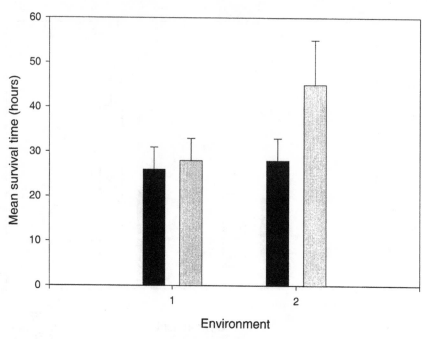

Figure 5.3. Hypothetical covariance of genes and environment

might be interpreted as indicative of an underlying genetic difference. For this reason, evolutionary geneticists assert genetic differences only when they can control the nature of environmental variation. Figure 5.3 shows how phenotypic differences can be misleading. In scenario 1, the two genotypes have a positive covariance with environment (that is, they share similar distribution over the two habitats). If we measured the mean survival times of the two organisms in scenario 1, we would conclude that the times are very similar. There would be no reason to suppose a genetic difference between them. However, in scenario 2, the two genotypes are unequally distributed over the benign and detrimental environments, and we would calculate a much larger mean for the genotype indicated by the gray bar, because it is differentially represented in the more permissive environment 2. Thus in the second scenario the covariance with respect to environment would exhibit a negative sign. (In chapter 10, I argue that this is precisely the situation that has always existed in attempts to measure intelligence between the so-called races).

Intelligence, however defined, is a phenotype; that is, it is an observable trait that arises from interactions between an individual's genes and his or her environment. Performances on standardized tests during the *Plessy v. Ferguson* era in the American South would not have been capable of revealing genetic differences in intelligence between the so-called races, even if such differences had existed. This is simply because standardized tests do not take environmental factors into consideration. "Blacks" and "whites" lived in different environments, one manifestly superior to the other in physical and biological infrastructure. Thus, the core error of all social Darwinian thinking results from the inability to separate genetic from environmental sources of variation in the phenotype.

Spencerism versus Darwinism

There are elements in Spencer's thinking that seem superficially to be in concert with Darwinian mechanisms of evolution. However, an informed analysis reveals that this is just appearance. Spencerism consists of the following components:

1. Progressive evolution of all things in the universe, including human society.
2. Malthusian population thinking: as the food supply increases arithmetically, population increases geometrically in relation to it; and the resulting shortage of food creates the struggle for existence.
3. A belief in "survival of the fittest," but more by the Lamarckian mechanism of the inheritance of acquired characteristics.
4. An attempt to provide a "scientific basis" for morality, such that European culture, particularly with its laissez-faire behavior, defined what was right.
5. An attempt to demonstrate that sociology could be categorized as a science. Spencer defined the task of sociology as uncovering the normal laws of social (progressive) evolution.
6. A belief that the driving force of progressive evolution was the conservation of energy.
7. A belief in an "unknowable" element in the universe that unified its direction and goals.

Darwin and, later, neo-Darwinians addressed these same issues, and it is instructive to compare the views of Spencerism and Darwinism point by point. The idea of progressive evolution dates to Lamarck, who thought that the Creator imbued all living things with an internal desire to improve. Some Darwinians believed in progressive evolution. This is the most elementary error of naive evolutionists. However, it is an error that even Darwin sometimes made, although it is clear that, in the main, Darwin realized that evolution was not always progressive. Darwin saw the action of natural selection through the differential reproduction of favored varieties; and if a loss of structural complexity met the needs of differential reproduction in an organism's environment, there was no reason such a loss could not occur. Darwin clearly describes the case of rudimentary organs lost because of disuse (for example, ancestral species of modern snakes once had legs), and he describes variations that seem to have no functional significance in *The Origin of Species*. In *The Descent of Man* he makes it clear that humans also have rudimentary organs, such as the appendix, and that much of human racial variation is of no functional significance. Thus Darwin saw that natural selection would mold organisms to their environments. If the environment changed, the direction of evolution would also.

Darwin recognized the essential soundness of Thomas Malthus's conclusion that there must be a struggle for existence. However, unlike Malthus, he provided a series of scientifically valid examples to demonstrate the fact that there must be natural checks other than food scarcity on the increase of all species. In a section of *The Origin of Species*, Darwin demonstrated the mathematics of exponential population increase. In his description of the checks to natural increase, he correctly identified the types of biotic and climatic sources of mortality that exist for all organisms. However, unlike Malthus, he recognized that the interactions between animals and plants that mediate the struggle for existence are highly complex. In a sense, Darwin set the stage for the modern discipline of ecology. Malthus and Spencer did not appreciate the implications of the complexity of ecological interactions in biological systems.

Spencer coined the phrase "survival of the fittest" to describe competition among human individuals and groups. Darwin even gave him credit for the phrase in *The Origin of Species*. However, Spencer argued that human progress resulted from the triumph of more-advanced individuals and cultures over their inferior competitors. Thus, he saw wealth and power as signs of inherent "fitness," and poverty as evidence of natural inferiority. Darwin's natural selection and Spencer's survival of the fittest were very different processes. The fatal flaw in the latter was the circularity of Spencer's definition. If fitness is defined by wealth, then by definition only the wealthy will survive. Thus, for Spencer, evolutionary worth could be determined only by post hoc methods. If one was poor, then that was the proof that one did not have the ability to succeed. As tautologies, such statements cannot be falsified.

Natural selection, however, is a testable hypothesis. Darwin described in detail how artificial selection in the form of breeding regimes had produced domestic varieties of plants and animals. This scheme could be tested a priori. That is, one could predict the organisms that should survive under a given set of conditions. The classic example is industrial melanism, studied by biologist H.B.D. Kettlewell from 1953 to 1955 in Birmingham, England. Kettlewell was able to demonstrate that pollution had changed the favored color of the moth population in Birmingham from white to black. The chief predators of these moths were birds that hunted them visually. In unpolluted forests the light moths prevailed because lichens were present to camouflage them. Pollution destroyed the lichens and darkened the bark, making the light moths easy prey for the birds. Today, there are thousands of experiments demonstrating the action of natural selection.

Another important difference between Darwin's and Spencer's thinking arises from the fact that they were describing fundamentally different modes of inheritance. Natural selection relied on a trait being genetically inheritable. Darwin was not aware of the true mode of heredity (discrete genes coded by DNA). However, his observations permitted him to formulate a mechanistic theory (blending inheritance accomplished by "gemmules" in the blood) that allowed natural selection to operate on traits determining biological fitness. Darwin's hereditary information was conserved in the sense that offspring resembled their parents, and this resemblance was truly constrained by the underlying principle of inheritance. Thus while Darwin's ideas concerning inheritance had major flaws, his theory of evolution did not depend upon the specific features of heredity.

In contrast, Spencer's description of society (and culture) dealt with a system where the inheritance of acquired characteristics did play a major role. Spencer's rules of inheritance in culture were Lamarckian (except that society is not always progressive). Thus, all people inherit their parents' genetic material, but they also inherit from them such things as culture, religion, education (or lack thereof), and general social status. For Spencer, the progress that he inferred resulted from the cultural, scientific, and technological achievements of the preceding generations. The error of Spencer and Galton (and their descendants) was that they assumed that these cultural traits were necessarily coded for by biological inheritance. Thus, in the late nineteenth and early twentieth centuries, Spencerians argued for unrestrained economic competition and against aid to the "unfit" poor. The theory was also used to justify racist and imperialist policies in Europe and the United States.

Darwin did not think that natural selection applied to morality. There was no direct transference through reproduction of the cultural qualities of humans. A female mouse consuming her litter in times of poor food availability cannot be accused of immoral infanticide. However, Darwin did argue that much of our behavior might have had roots in natural or sexual selection. Certainly Darwin did not believe that the harsh conditions of the poor in human society were directly the result of natural selection. Although he did think that intellect was a highly inheritable trait, it did not necessarily guarantee an individual's fitness. Neither did he think that intellect necessarily translated into wealth. As we have seen, he did feel that civilized nations had advantages in competition with barbarous ones. Conversely, he also thought that natural selection had produced altruistic capacities in our species. Personally he opposed slavery, and he uttered the famous statement, "If the misery of our poor be caused not by the laws of nature, but by our institutions, great is our sin."[6]

The fact that Darwin discussed the importance of evolution in influencing the origin of the human condition meant that he must have thought that social conditions were not "random." He was not, however, concerned with elucidating the specific relationships between the biological evolution of humans and their social forms. This development would have to await the rise of sociobiology, of which E. O. Wilson's *Sociobiology*, published in 1975, was the opening manifesto. Spencer was by no means the first sociologist attempting to develop general laws of social change. At least five years before Spencer, Marx had formulated historical materialism, a theory of progressive social evolution toward democracy. To Marx, society's evolution was driven by changes in the modes of production, which were also natural. These changes were driven predominantly by economic, not biological, forces. Marx did think, however, that human economics arose from innate biological and psychological sources. Marx, and especially Engels, also utilized Darwin's ideas, believing that the transition between apelike ancestors and modern humans was important for the formulation of social theories. They differed with Spencer in that they did not think that laissez-faire capitalism defined the pinnacle of human evolution. Capitalism was an advance over feudalism, but it was a stage on the way to socialism and communism, the stages of society that would allow the true expression of human potential.

Because Darwin did not generally think that evolution was progressive, the idea of the conservation of energy driving the progression of organic forms did not enter his thoughts. Spencer did hit on the correct idea, in a roundabout way, in that living cells (and therefore organisms) do obey the laws of chemistry. Natural selection usually favors organisms that efficiently partition their energy budget into differential reproductive success. Such partitioning can, however, just as easily lead to "retrograde" evolution. For example, many animals utilize the mode of parasitism to survive and reproduce. In the Schistosome flatworms, all male structures have been reduced to a bare minimum. Males essentially live out their lives attached to females, with only a functional testis. Nervous, muscular, and digestive systems are all reduced. The worms reside in the bloodstream of their human host, absorbing predigested materials directly through their integument.

Finally, Darwin did not think there was a central "unknowable" force driving the universe. For him, natural selection was sufficient to account for all the organic complexity of the earth.

It is clear that Spencerism and Darwinism are different theories. Tests have demonstrated that natural selection and evolution are valid descriptors of the general course of organic life, and Spencer's views on society, biology, and progress in the universe have been proven false. Unfortunately, this demolition did not happen until social Darwinism and eugenics had developed to the point where they could cause considerable harm.

CHAPTER 6

Pseudoscience and the Founding of Eugenics

Pseudoscience is a pernicious feature in the history of the study of nature, and the line between science and pseudoscience is often unclear and fluid. Pseudoscientific theories and methods, which borrow only the superficial qualities of scientific investigation, are sometimes intentionally passed off as scientific, but they can also arise when investigators fail to properly articulate the research question they wish to pursue. In the latter case, questionable ideas may later evolve into proper scientific theories. For example, the idea of continental drift, first proposed in 1912 by Alfred Wegener (1880–1930), was originally based on pseudoscientific premises. Because the coastlines of South America and Africa looked as though they could fit together like puzzle pieces, and because of other circumstantial evidence, Wegener concluded that perhaps the continents had once been connected and had later drifted apart. In Wegener's time, however, the means for testing this hypothesis did not exist. The testability of hypotheses is a crucial component that distinguishes scientific hypotheses from pseudoscientific speculation. Most importantly, it is the desire of researchers to formulate their hypotheses in ways that allow critical testing that differentiates science from pseudoscience. In the 1960s, with the measurement of magnetic pool orientations and the dating of the oceanic floor, Wegener's ideas were verified. What had begun as the pseudoscientific assertion of continental drift was later incorporated into the scientifically valid theory of plate tectonics.

Pseudoscience may also result from conscious deception, fraud, or a psychological predisposition to believe certain outcomes. We have already seen in chapter 3 an example of the latter: the polygenists' use of craniometry to infer the worth of human beings. Although Samuel Morton's studies showed no conscious bias, his work was, in the end, biased by a series of systematically recurring mistakes in data collection and interpretation. An even grander mistake present in Morton's research and craniometry in general was their failure to establish a direct quantitative relationship between head size and intelligence. Neither did craniometry establish a causal mechanism by which the absolute size of the brain was responsible for differential intellectual performance. It relied instead on an examination of how the ratio between body size and head size differed from species to species. Thus the research program suggested that because mice and rats have larger ratios of brain size to body size than frogs, and because mice and rats clearly exhibit more-complex behavior than frogs, the relationship between intelligence and this ratio must hold true within species also.

However, when unbiased measurements of human brains were made, there was no evidence for differences in brain size. In 1838, Friedrich Tiedemann, an anatomist from the University of Heidelberg, measured the brains from fifty cadavers (both Negro and European) and found no weight differences. In 1900, the French anatomist Joseph Deniker collected data on 1,100 brains of Negroes and Europeans and found no discernible difference. Finally, Franklin P. Mall, from Johns Hopkins, utilizing a sample that had been used earlier to profess a difference in the brains of Negroes and Europeans, remeasured the brains using a blind identification technique and also found no difference. The twentieth-century anthropologist Ashley Montagu concluded that the average cranial capacity difference between blacks and whites was about 50 cubic centimeters.

However, all morphological features in humans show considerably more variation *within* than *between* the so-called racial groups. For example, the Kaffirs and Anaxosa of Africa have larger crania than do whites, as do the Japanese, Eskimos, and Polynesians, a fact that is particularly notable when one recognizes that the Eskimos and Polynesians did not score high by European standards of civilization. In addition, even the cranial capacities of Neanderthals were greater than those of modern Europeans.[1] It is interesting to note in the light of this model that the large-brained Neanderthals may not have left any descent lineages in modern humans.[2] Even Paul Broca would admit in the end that the differences in cranial structure, and therefore brain size, within a race were often greater than the differences among the races, making craniometry of dubious value. Thus, the polygenist research program relied on a false assumption.

The practice of building whole theoretical constructs on false or untestable assumptions is a hallmark of pseudoscience, which is often associated with vested social agendas. The results of the polygenists were definitely consistent with the views of the slaveholding society. We have asked to what degree the desire to validate the institution of slavery motivated polygenist research. If the world of 1850 had not been one of extreme social disparity, would anyone have been interested in the relationship between racial variation and head-size variation? The history of Western concepts of race has straddled the line between poor science and pseudoscience. No thinkers illustrate this process more vividly than Joseph Arthur Comte de Gobineau and Sir Francis Galton.

Gobineau, Race, and Civilization

Four years before the publication of *The Origin of Species*, the Comte de Gobineau (1816–1882) would author his *Essay on the Inequality of Human Races* (1853–1855). This work stated that the Aryan, or white, race was superior to all other races. Gobineau was an aristocratic dilettante who wrote novels, as well as books on religion, philosophy, and history. During his studies, he became concerned with the problem of why great civilizations seemed destined to decline. He concluded that the source of the decay was their racial composition. His reasoning was as follows: the first step in the formation of a nation was the uniting of several tribes, either by alliance or conquest. Certain ethnic taxa seemed to be capable of this process, whereas others failed. He did not locate this inability in external features, such as climate, reasoning that great nations had been found in a variety of climates, cold, temperate, hot, inland, and coastal. The mechanism of alliance within these nations was intermarriage. As the nations grew and became international, strangers would be drawn to them. If these strangers were racially inharmonious, admixture would result in degeneration of the society. Gobineau's theories of the deleterious effect of racial admixture would become the ideological basis for later policies of immigration restriction in Aryan nations.

Despite his negative views of most races, Gobineau did not believe in a simple stereotyping of all individuals within a race. He was a monogenist and rejected Benjamin Franklin's view that Negroes were animals who ate as much as possible and worked as little as possible. Gobineau's views are more similar to those of modern theorists who focus on the average difference in intelligence or ability between races.

Gobineau's metric of a race's worth was its ability to found a "great civilization." He listed seven Caucasoid civilizations in the Old World (the East Indian, Egyptian, Assyrian, Greek, Chinese, Roman, and Nordic/Aryan), three Mongoloid civilizations in the New World (Alleghanien, Mexican, and Peruvian). The Alleghaniens were supposedly an ancient mound-building culture that inhabited North America before the coming of the Europeans. Gobineau denied that any Negroid race had ever produced a great civilization (see table 6.1).

The view that there were no great Negro civilizations was held by many

Table 6.1. Gobineau's Great Civilizations, by Racial Group

Negroid	Caucasoid	Mongoloid
None	East Indian	Chinese
	Egyptian[a]	Alleghanien
	Assyrian	Mexican
	Greek	Peruvian
	Roman	
	Nordic/Aryan	

SOURCE: Gobineau's views on civilization are summarized from J. R. Baker, *Race* (New York: Oxford University Press, 1974), 37.
NOTE: These groups have been arranged into modern racial categories, which Gobineau and nineteenth-century anthropology also used.
[a]The historians and anthropologists of the nineteenth century denied any African or Negroid influence in Egyptian civilization. Furthermore, Gobineau saw all these racial categories as distinct and not the result of hybridization.

of Gobineau's ideological descendants. A perfect example of the denial of African ability to produce a civilization is the treatment of the Shona city known as Great Zimbabwe.[3] Modern archaeological work suggests that the city was well established as a trading center by A.D. 1400. Artifacts have been found there originating from Syria, China (Ming Dynasty), and Persia. When the German explorer Karl Mauch arrived in 1871, he concluded that this city was the mythical Ophir of the story of Solomon. Mauch believed that the site (particularly the architecture and stone cutting) was too sophisticated to have been produced by Africans and that it must have been the work of Phoenician or Israelite settlers. A series of European "archaeologists" who followed Mauch to the city looted it and destroyed valuable stratigraphy surrounding many structures, thus making a complete evaluation of Great Zimbabwe difficult.

David Randall-MacIver would be the first European (1905) to recognize that Great Zimbabwe was contiguous with the culture of the Shona people, who still inhabited the region. These findings were censored by the white supremacist Rhodesian Front (established by former Prime Minister Ian Smith) from 1965 until independence in 1980. This treatment of archaeological finds in sub-Saharan Africa was typical of nineteenth- and early-twentieth-century European anthropologists. Artifacts found in Old Stone Age African cultures would be labeled as New Stone Age, giving the impression that African cultures arrived at various technologies and practices uniformly later than did Asian or European cultures.

For Gobineau, racial hybridization was the root cause of the decline of great civilizations. He thought that the decline of superior Jewish society resulted from its mixture with inferior Hamitic tribes (remember Ham had supposedly been the originator of the "black" and "yellow" races). This idea contains the seed of dysgenic theories, which suggest that greater reproduction or immigration of genetically inferior populations will result in the overall decline of a nation. Dysgenics, the study of racial degeneration, is essentially a corollary of eugenics, the science of improving the hereditary qualities of a race. Dysgenics would be further developed by Francis Galton. It would also become the rationale for the early-twentieth-century immigration restrictions in the United States. It is also now clear why Gobineau's theories of racial superiority later influenced Nazism. Great civilizations could prosper only so long as they remained racially pure. The Jews would be considered racial impurities infesting the Germanic nation. The German biologist Ernst Haeckel, the composer Richard Wagner, and Wagner's son-in-law, Houston Stewart Chamberlain, were devotees of Gobineau's ideas. These individuals, in turn, had an important influence on Nazi race theorists, including Adolf Hitler.

Most modern anthropologists consider Gobineau's thinking worthless, precisely because external factors that he discounted do have a major influence on the growth and development of civilizations. Modern-day biologist Jared Diamond describes precisely how the contingencies of nature—climate, fauna, and flora—have a profound impact on the types of civilizations that can develop in a region.[4] For example, much of the continent of Africa suffers from high rainfall levels. This rainfall leaches nutrients from the soil, producing soils that are infertile. Thus, the paradox of tropical rain forest environments is that their nutrients are locked up in the existing biomass in the form of trees. The clearing of forests leads to even greater loss of soil nutrients. Such nutrient-deficient environments preclude the style of agriculture practiced in temperate climates. Stable agriculture is the prerequisite for stationary populations, and the growth of population is, in turn, a prerequisite for industrialization.

Climates also change over time. One reason for the decline of stable populations in the Magrib was the increasing desertification of northern Africa, caused by long-term global climatic changes and accelerated by human agricultural activity in the region.

In addition to climate, the nature of the flora in a particular region affects human populations. Some types of plants are more desirable for human consumption than others because of differences in their chemical makeup. Many of these differences are related not to human activity but to insect activity. There are more insects in the tropics (because of the lack of cold winters to reduce their numbers), and thus plants respond by producing greater amounts of naturally insecticidal compounds, which in turn determine the palatability and toxicity of these plants to humans.

These chemically complex natural compounds include alkaloids, narcotics, tannins, and terpenoids. Many notable drug compounds found in plants (such as aspirin, cocaine, opium, quinine, and others) are derived from chemicals found in tropical species.

The fauna of a region also affects the growth of societies in that region. Consider, for example, the horse, which was present in Europe and Asia Minor but absent in North America and tropical Africa. Horses were crucial to the conquest of Mexico by Cortez. Diseases caused by parasites are more devastating in the tropics. It has been shown that the combination of mosquitoes, nematodes from the *Onchocerca* and *Schistosoma* genera, and other pests makes cultivation of much of the land of tropical Africa impossible. Onchocerciasis is particularly damaging to draft animals, and the absence of oxen and horses forced tropical African societies to rely on the hoe rather than the plow. Furthermore, populations that carry high worm burdens utilize most of their daily caloric input to maintain normal temperatures. For many reasons, the temperate zones were more compatible with the growth of high densities of human populations than the tropics.

These points had been recognized as early as 1830. Alexis de Tocqueville (to whom Gobineau served as a secretary in 1849) recognized that European agricultural and hunting activities were having a negative impact on the populations of American Indians. Tocqueville reasoned that the hunter-gatherer existence of the Indians was tied to the forests and herds of animals that had been abundant before the coming of the European. The loss of these resources doomed the indigenous tribes. In addition, the subsistence or even starvation conditions of the American Indians would have made them more susceptible to disease (including smallpox). The interaction of disease prevalence and nutrition is another example of gene and environment interaction. It is suggested that the founder populations that gave rise to the American Indian populations of the New World may never have been exposed to smallpox. Thus, mutations conferring resistance to this and other parasitic diseases borne by Europeans and Africans to the New World would have been rare. Some anthropologists suggest that these diseases may have killed more Indians than did wars.

Gobineau's views of the centrality of race in civilization fit well within the framework of pseudoscience. He began his investigation convinced that his ideas were correct. To test his ideas, he formulated hypotheses that were wholly correlative and based on highly subjective criteria. One could have asked whether any evidence could possibly have disproved Gobineau's hypothesis that certain races were not capable of creating great civilizations. He had essentially produced a test of the degree to which a culture matched European expectations. For some, he would readily identify "greatness." Nongreat races did not match his criteria simply because he refused to see the great attributes of their civilizations. He did not recognize the accomplishments of the civilizations of Ethiopia, Nubia-Meroe, Songhay, or Ghana, all of which were known in his time. He did not adequately propose or investigate alternative theories that might have explained his observations. For example, he did not consider the possibility that great empires declined because they became too large to be effectively governed. A great empire that acquired large amounts of territory would also by coincidence incorporate a number of races and cultures. Thus, his proposals were tautological; they were true by definition and by criteria that could be molded to yield the conclusion desired. Finally, Gobineau also fell into the great trap of racist pseudoscience, the assumption that cultural behavioral traits are genetically determined. The science of this period had no workable theory of heredity, which fostered pseudoscientific speculation concerning how heredity determined racial character.

Sir Francis Galton, Founder of Psychometry

Writing about the lower intellect of the English settlers of the United States, Canada, and Australia, Francis Galton stated, "Better class Englishmen prefer to live in the high intellectual and moral atmosphere of the more intelligent circles of English society, to the self-banishment among people of altogether lower grades of mind and interest. . . . England has certainly got rid of a great deal of refuse, through means of emigration"[5] Galton founded the discipline of psychometry (the examination of variation in human intelligence), although he was best known for his work in anthropology and heredity. He also made some significant forensic discoveries, including the importance of fingerprints as individual identifiers. Galton was Charles Darwin's cousin and a member of the X Club and the Ethnological Society, and he has been described as a Victorian genius. In this I concur, in that the real genius of the Victorian ruling class was its ideological domination and exploitation of the common people. Galton was also the founder of eugenics.

Galton was born near Birmingham, England, and was educated at King's College, London, and Trinity College, University of Cambridge. He traveled in Africa in 1844 and 1850 and subsequently wrote *Narrative of an Explorer in Tropical South Africa* (1853) and *Art of Travel* (1855). Galton became interested in heredity and the measurement of human physical characteristics in the years after his return to England, perhaps because of his own sterility and mental problems (he suffered a mental breakdown in 1866, which may have been due to obsessive-compulsive personality disor-

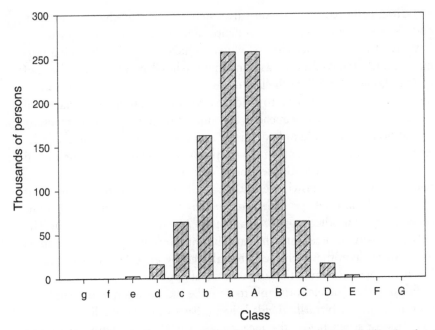

Figure 6.1. Quételet's principle. The figure illustrates the frequency distribution for a hypothetical population of 1 million people measured for some trait (G, g = 0.01/1000; F, f = 0.2/1000). Data recalculated from Francis Galton's *Hereditary Genius* (1869), after J. R. Baker, *Race* (New York: Oxford University Press, 1974).

der). None of Galton's brothers had children either, and we know that Galton's cousin Charles Darwin had sickly children and suffered from panic disorder. Galton was the son of Violetta Darwin, Charles's aunt, and both Charles and Francis received one-fourth of their genetic material from Erasmus Darwin. It may seem that this might prove Galton's claim of the hereditary nature of genius. However, it must be remembered that both Charles and Francis were raised in an environment of considerable wealth. And Galton's psychological condition may have explained his great ability to take careful, almost obsessive, observations. Galton may serve as a better example of the heritability of psychological disorders than of the heritability of genius, although there is some evidence that people who suffer from these problems are often highly creative.

Galton's analytical approach to his study was stimulated by the earlier work of Belgian statistician Adolphe Quételet. Quételet had defined a principle that he called the law of the deviation from an average. What Quételet was describing with this law was the normal probability distribution, first defined by the great mathematician Carl Friedrich Gauss (1777–1855). Figure 6.1 shows a hypothetical example of Quételet's principle.

Galton collected statistics on various characteristics—height, strength, and so on—for a large number of people and found that frequency distributions of these characteristics followed Quételet's principle, being normally distributed. Galton's reasoning that the intellectual characteristics of people should also follow the law of deviation from the average is probably his greatest contribution to modern psychology. But his explanation of the pattern and its consequences for human society was fundamentally flawed because he could only observe particular patterns; he could not determine the causative agents that were producing them. Galton committed the main logical error in all pseudoscientific investigation. Despite the fact that he had no real knowledge of the causal mechanisms producing what he observed, he used the observed patterns to make further predictions concerning what other types of related patterns should exist in nature. The only valid way to determine the causality of particular patterns in data is to construct hypotheses that can be falsified by experiment. This allows for the sequential elimination of conjectural hypotheses.

In using the Quételet principle to describe the variation in human intelligence, Galton arbitrarily divided the grades of human intellect into sixteen classes: seven below the mean (classes g–a), seven above the mean (classes A–G), and two extreme classes (X for truly outstanding individuals and x for true imbeciles). He concluded that 1 person in 79,000 should fall into grade G (and conversely g). Galton ranked the intelligence of dogs and varieties of human beings on the basis of this scheme. Note that he had no objective means of measuring the "intelligence" of these groups; the Binet intelligence test, for example, would not be invented until 1905. His ranking is entirely the result of his subjective impressions of the intelligence of these groups. Table 6.2 shows how, in *Hereditary Genius* (1869), Galton arranged these groups according to intelligence classes.

His arrangement is not credible. Galton concludes that the highest cognitive class of dog intellect (X) is equivalent to or superior to classes x–a of Australian aborigines, classes x–b of Negroes, classes x–d of Englishmen, and classes x–e of the ancient Greeks! This conclusion is amazing, particularly given the fact that Galton's travels must have shown him that virtually all Australoids and Negroes possessed language and tool-making abilities that all dogs lack. Certainly, his servants displayed at least these skills. It is even harder to believe that Galton did not recognize language and tool making as important cognitive skills that should have been included in any definition of human intelligence. His failure to recognize this is compounded by his even greater error of attempting to compare, subjectively, the intellectual abilities of organisms in different species or of populations within the same species that had lived in different historical

Table 6.2. Galton's Comparison of the Cognitive Abilities of Human Races and Dogs

Group	Cognitive class															
Greeks (450 B.C.)	x	g	f	e	d	c	b	a	A	B	C	D	E	F	G	X
Englishmen		g	f	e	d	c	b	a	A	B	C	D	E	F	G	X
Negroes				e	d	c	b	a	A	B	C	D	E	F	G	X
Australians					d	c	b	a	A	B	C	D	E	F	G	X
Dogs													E	F	G	X

SOURCE: F. Galton, *Hereditary Genius: An Inquiry into Its Laws and Consequences* (London: Macmillan, 1892).

NOTE: Galton felt that cognitive abilities were distributed according to Quételet's principle. On his scale, the different races overlapped in their cognitive abilities, the lowest humans being equivalent in intelligence to some breeds of dogs.

periods. Galton certainly understood that intelligence, however defined, was potentially subject to natural selection, but the behavioral traits that are conducive to differential reproduction and survival of dogs cannot be validly compared to the traits that lead to the same in humans. Any such comparison is, of course, made even more suspect by the fact that dogs, unlike humans, had been subjected to severe directed breeding.

Galton may have been biased toward inferring hereditary differences in intellectual performance between groups by the fact that various breeds of dogs have different personalities and differ in their ability to learn. These hereditary differences between dog breeds have often been used to support the idea that natural selection must have produced differences in the intellectual characteristics of the human races. However, the analogy between dogs and humans is a false one. In *The Origin of Species* Darwin discussed how animal breeds were the products of conscious human selection (artificial selection), bred for the requirements of human activities. Thus, a dog like the dachshund may have had its short legs selected so that it could pursue small rodents into burrows. This trait might be beneficial to the human breeders, but it is of doubtful value to the dog. Indeed, animals that are selected for aggression, such as the Doberman and the pit bull, suffer from a number of problems that impact their overall fitness.

The type of intense artificial selection used to differentiate traits in animal breeds is not seen in humans. My work with fruit flies provides an example of the intensity differential between artificial and natural selection. I have evaluated quantitative traits related to overall fitness in fruit fly stocks that have undergone selection for life-span characteristics. When a 10 percent selection differential per generation was used, these stocks exhibited new life-spans in less than sixty generations. Thus a 10 percent selection differential showed a rapid change in a polygenic trait.[6]

Like fruit fly life-spans, human skin coloration is also a polygenic trait, determined by about six to eight genes.[7] Northeastern Asian populations from the Old World arrived in the New World about thirty-five thousand years ago (based on evidence from archaeological sites in New Mexico and South America). In the New World there is no pronounced skin color gradient like that observed in the habitats of the Old World. If we set the average human generation as fifteen years in length, then the time to complete a generation cycle is thirty years (female parent grows to reproductive age of fifteen, then rears her daughter to reproductive age of fifteen). Using this figure, we can calculate that 1,166 generations have passed without a return to the original melanic skin compositions of the original *Homo sapiens* of the Old World tropical zones. This number indicates that the selection favoring differences in skin pigmentation is extremely weak. In fact, many selection differentials impacting other traits associated with racial variation in human populations are also weak.

Owing to intense selection and small population stocks, domestic animals, like dogs, also suffer from inbreeding effects. Inbreeding tends to make the genome of a stock homozygous at many loci, which has the overall effect of lowering the fitness of a given line, even if that line is fixed for some desirable trait. Inbred stocks will differ from one another in terms of the specific fitness defects. The only reliable or interpretable result from the differences shown in personality or intellectual characteristics of domestic breeds is that genetic variation for these traits does exist in the species in question. Human populations have occasionally undergone population bottlenecks that have produced some inbreeding, but not at the level experienced in artificial selection. Therefore, the idea that human races are differentiated in genes associated with intelligence because of regional differences in natural selection is theoretically and empirically ill-founded.

Galton and the Negroes

Francis Galton discussed his views on the intellectual inferiority of the Negroid race openly: "*Fourthly, the number among the negroes of those whom we should call half-witted men, is very large. Every book alluding to negro servants in America is full of instances. I was myself much impressed by this fact during my travels in Africa. The mistakes the negroes made in their own matters, were so childish, stupid, and simpleton-like, as frequently to make me ashamed of my own species* [italics added]."[8]

Comments of this nature may explain why Galton thought that various breeds of dogs were higher in intellect than some races of humans. In his zeal to classify Africans as an inferior type, Galton failed to note or understand that most of the literature on American slave intelligence was written by slaveholders. The fact that the entire way of life of these slaveholders rested on a presumption of African inferiority surely influenced their depiction of blacks. Numerous sources have described how acting the part of the buffoon was integral to slave survival and well-being. On the other hand, Galton classified François Toussaint Louverture, the leader of the successful 1801 slave rebellion in Haiti, as Negro class X, equivalent to his intelligence class F in the English (1 in 4,300). However, it must be remembered that two English grades existed above F: G (1 in 79,000) and X (1 in 1,000,000). The Afro-Caribbean Toussaint Louverture had to earn his X ranking by defeating the best English, French, and Spanish generals sent against him—all men whom Galton would have ranked in English class G! Galton dismissed the social difficulties of Negroes in the United States, and this allowed him to locate their average intelligence two grades below that of Englishmen (see table 6.2).

Another European observer of racial conditions in the United States around this time, Alexis de Tocqueville, came to a conclusion opposite from Galton's: "A natural prejudice leads a man to scorn anybody who has been his inferior, long after they have become his equal; the real inequality due to fortune or law, is always followed by an imagined inequality rooted in mores; but with the ancients this secondary effect of slavery had a time limit, for the freedman was so completely like the man born free that it was soon impossible to distinguish them." Tocqueville also went on to observe that in the United States "race prejudice seemed stronger in those states that had abolished slavery than in those where slavery still exists, and nowhere is it more intolerant than in those states where slavery was never known."[9]

Galton, however, did not seem to observe the powerful impact of racist colonial society on the behavior of African Americans or Africans. He boldly makes the following statement to support his racist characterization of intellectual types:

A traveller in wild countries also fills, to a certain degree, the position of commander, and has to confront native chiefs at every inhabited place. . . . It is seldom that we hear of a white traveller meeting with a black chief whom he feels to be the better man. I have often discussed this subject with competent persons, and can only recall a few cases of the inferiority of the white man—certainly not more than might be ascribed to an average intellectual difference of three grades, one of which may be due to relative demerits of native education, the remaining two to a difference in natural [i.e., hereditary] gifts.[10]

The idea never seemed to cross Galton's mind that the survival and well-being of African Americans or colonized Africans might have depended on their giving the appearance of intellectual inferiority. In addition, Galton's cultural chauvinism made it impossible for Africans to provide any evidence that would have convinced him of their intellectual worth.

Galton, Class, and Genetic Endowment

Galton claimed that the great achievements of England were made by a few English families. He arrived at this conclusion by examining obituaries in the English newspapers, reasoning that only great people earned obituary treatment, and that the number of great people relative to the general English population was 1 in 4,000. He found that many of those who had been given obituaries were descended from or related to other subjects of obituaries. Galton made the logical error of not reporting negative instances: that is, that there were many Englishmen of accomplishment who did not arise from famous families. In addition there were even more offspring of famous families that did not achieve Galton's idea of greatness. It is trivial to show that the number of English historical figures whose offspring and close relatives were not distinguished is much higher than the number of people who fit Galton's hypothesis. Galton simply chose the examples that supported his thesis and ignored those that contradicted it, a classic methodological error among psychometricians. Certainly if one examines the royal houses of Europe, many of whose members have suffered from inbreeding, one finds little evidence of "hereditary genius." Witness, for example, the madness of King George III due to the genetic disease porphyria.

Galton also raised the specter of dysgenesis. Intellectual dysgenesis, for Galton, resulted from the less intelligent classes reproducing at rates higher than the more intelligent classes. If this were to continue, the result would be a dilution of intellectual talent leading to the breakdown of social institutions, according to Galton. Dilution has been a consistent concern of the psychometric movement. Galton's solution to dysgenesis was to encourage the most intelligent classes in English society to reproduce in greater numbers, a so-called positive eugenic program. Conversely, preventing the reproduction of genetically inferior classes, "negative" eugenics, should have the same effect. Lewis Terman, of Stanford University, an American disciple of Galton and the originator of the eugenic version of the intelligence test, would describe the danger of dysgenesis directly:

Their dullness seems to be racial, or at least inherent in the family stocks [i.e., strains of given races] from which they come. The fact that one meets this type with such extraordinary frequency among Indians, Mexicans, and Negroes suggests quite forcibly that the whole question of racial differences in mental traits will have to be taken up anew and by experimental methods. The writer predicts that when this is done, there will be discovered enormously significant racial differences in general intelligence, differences which cannot be wiped out by any scheme of mental culture.

Children of this group should be segregated in special classes and be given instruction which is concrete and practical. They cannot master abstractions, but they can often be made efficient workers, able to look out for themselves. *There is no possibility at present of convincing society that they should not be allowed to reproduce, although from a eugenic point of view they constitute a grave problem because of their unusually prolific breeding* [italics added].[11]

In a 1901 lecture to the Huxley Institute, Galton claimed that the "brains of our nation lie in the higher of our classes." Galton wished type X children to be identified at birth and stated that if the nation were to pay £1,000 per head to their parents, it would be a bargain. He reasoned that such children would found great industries, amass great fortunes, and thus increase their own wealth and that of the multitude. Galton suggested that class X women should bear at least one extra male and female child, in addition to those they would normally have produced.

An examination of Galtonism in hindsight is revealing. The first question of interest is why Galton's views received so much attention in his own time and subsequently. Galton was knighted in 1909, and modern psychometricians hail him as the founder of their discipline. Yet if Newton or Darwin had presented such flimsy evidence to support his theories, those theories never would have been accepted, let alone have survived to command devoted followers today. Galton's eugenics devotees in following generations even went so far as to compare his IQ with that of his cousin Charles Darwin. In 1917 Lewis Terman calculated the IQ of his dead hero Galton from materials such as test scores and subjective evaluations of contemporaries and concluded that Galton had a score of 200 (only 1 in 50,000 have that score), as compared to 135 for Charles Darwin. The IQs of other dead figures were calculated, and among those who exhibited scores lower than Galton's were Copernicus, Cervantes, Faraday, Harvey, and Lavoisier!

We now know that Galton's reasoning from the Quételet principle to human physical and mental characteristics did not take into account environmental effects on gene action. Virtually none of Galton's explanations are valid when viewed from the standpoint of modern quantitative genetics, though Galton's students, Karl Pearson, for example, would later develop statistical tools that we currently use to examine such quantitative characters.

The scientific accomplishments of Galton and his followers were small in importance relative to their social impact. Galton's ideas led to significant eugenic movements in the United States, Latin America, and Europe. One consequence of this development was the involuntary sterilization of thousands of people under the direction of eugenic principles. Certainly eugenics has to take some of the responsibility for the Holocaust. Galton's scientific accomplishments are sufficient for some to still consider him an intellectual hero. Whereas for others (this author included) he was an intellectual mediocrity, a sham, and a villain. We still cannot effectively determine the sources of variation in human intellectual performance, no matter how measured. Hence, all claims about genetically determined intellectual differences among races are at present necessarily pseudoscientific.

Eugenics and Immigration to the United States

The champions of Anglo-Saxonism in the United States established eugenics as a political tool between the end of the Civil War and the beginning of the twentieth century. They declared that the Anglo-Saxon population should be credited with all positive developments in Western civilization. Democracy itself was to be seen as a unique development of the Teutonic races. Other races had copied it, but it could flourish only among people of the Anglo-Saxon race. Dilution of the America population by non-Teutonic races threatened to destroy American social institutions. The young Woodrow Wilson would comment on the biological threat to Anglo-Saxon institutions created by the influx of non-Teutonic races. These people he described as coming "out of the ranks where there was neither skill nor energy nor any initiative of quick intelligence."[12] Teddy Roosevelt would campaign against birth control among Anglo-Saxons because it diminished the birthrate of the native American stock. Of course, Roosevelt was referring to those Americans of Anglo-Saxon heritage when he used the term "native American stock."

Before the 1850s, European immigration to the United States had generally been welcomed; there were still abundant opportunities in farming and industry. In this period the majority of European immigrants were from the so-called Teutonic nations (Germany, Austria, the Netherlands, England, and so on). The resolution of the political and social issues created by the Civil War allowed the rapid expansion of industry in the northern states. African Americans, still located mainly in the South, were not considered a viable industrial labor force, and the advent of sharecropping created an incentive for southern landholders to prevent African American migration to the North or the West. On the contrary, southern landowners attempted to stimulate the flow of immigrants into the southern peonage system, including Chinese laborers from the West and non-Teutonic immigrants from Europe. The possibility of the inflow of new labor to the South prompted Booker T. Washington to address this issue in his famous "Atlanta Compromise" speech of 1900, in which he warned southern whites not to look to Europe to provide the labor that they could easily find from the Negroes already living there.

American anti-immigrationism developed most strongly in New England, because of a series of coincidental biological and social changes that contributed to a demographic shift impacting the old Anglo-Saxon families. The soils of this region had been intensively cultivated since the seventeenth century, and the abundance and availability of better farmland in the West stimulated the migration of young men from the old New England families out of the region. Simultaneously, industrial demand for labor was increasing, and the immigrants now answering the call were mainly non-Teutonic (Italians, Jews, Poles, Serbs, Hungarians, Greeks, and others). An 1883 study found that in this region the native New England birthrate had declined and that that of the immigrants was increasing. Also, near the end of the decade, an unidentified epidemic disease decimated many native New Englanders while leaving many of the immigrants relatively untouched. These two events led to cries of "biological" decay among the native stocks.

The population dynamics in New England serve as a perfect example of how people fail to separate the biological, social, and cultural factors affecting population changes. The New England population was affected by all three factors. For example, the higher migration rate of the native-born population to the West is readily explainable socially. These individuals would have had both the means and the knowledge of American conditions to facilitate their migration westward. In addition, the higher birthrates in the immigrant communities can be readily explained by their impoverished state. We know that inequitable distribution of wealth is a primary contributor to the greater birthrate of the poor classes. Culturally, higher birthrates might have been expected in the Catholic immigrant groups, owing to potential differences in birth control practices. Finally, the differential impact of an epidemic disease could have resulted from the fact that the older New England families were most likely genetically related. A pathogen to which these families had a diminished immune response would have had greater impact on them than on a group of immigrant families exhibiting wider genetic diversity. No where is there any evidence of the intrinsic biological decay of the native New England families.

The 1870s and 1880s saw violent outbursts of nativism and anti-immigrant hysteria. These were accompanied by anti-Catholicism and anti-Semitism (many of the new immigrant populations were Catholic or Jewish). Popular literature such as the works of Jack London and the international Teutonist Rudyard Kipling championed the virtues of the Anglo-Saxon. Immigrants were vilified, particularly as the originators of labor union strife and as radicals, anarchists, and communists. The Haymarket Square bombing in Chicago in 1886 was tied to foreigners. Popular Teutonist scholars like the sociologist R. A. Ross and Oxford University's E. A. Freeman warned Americans of the racial pollution of the Aryan/Anglo-Saxon bloodline by the Negroes, the Irish, and the Jews. These arguments were beginning to resonate in the minds of the United States' so-called native-born.

In 1894, three members of the Harvard class of 1889—Prescott F. Hall, Robert DeCourcy Ward, and Charles Warren—founded the Immigration Restriction League (IRL). A year later, the manifesto of the new anti-immigration movement would be published in the *Atlantic Monthly* by General Francis Walker, former chief of the Bureau of Statistics and superintendent of the census. Walker, who was strongly influenced by social Darwinism and by Gobineau's theories on the inequality of race, articulated two great fears in his article entitled "Restriction of Immigration": that the indigenous "pauper classes" would out-reproduce the Anglo/Nordic stocks and that immigration from non-Nordic countries would dilute the United States' pure blood. Practically, it was easier to control the latter threat than the former. Walker claimed that the American population in the years 1830–1880 would have grown at the same rate with or without European immigration, and he suggested that the native-born population had stopped reproducing in this period because of its abhorrence of the foreign-born races that could not be assimilated into the United States. He thus described the new immigrant stocks: "They are beaten men from beaten races; representing the worst failures in the struggle for existence. . . . They have none of the ideas and aptitudes which fit them to take up readily and easily the problem of self-care and self-government, such as belong to those who are descended from the tribes that met under the oak-trees of old Germany to make laws and choose chieftains."[13]

Within ten years the IRL would include among its members the presidents of Harvard University, Bowdoin College, Stanford University, Western Reserve College, Georgia Tech, and the Wharton School of Finance. It was the IRL that devised the literacy test for immigrants that it introduced to Congress through former Harvard history professor and then senator Henry Cabot Lodge. The IRL also maintained a permanent lobby in Washington. By 1924 it had succeeded in convincing Congress of the biological inferiority of non-Nordic races. These sentiments would be incorporated into the Immigration Act of 1924. The act, which regulated European immigration based on the census of 1890, was favorable to the Nordic countries. The act gave each country of origin a rate of 2 percent of the total immigrant pool.

Although European immigration had been encouraged prior to the 1880s, Chinese immigration had not. For example, the annual report of the secretary of the treasury shows that from 1894 to 1901 only 84 percent of the 37,374 Chinese who applied were admitted to the United States. This figure should be compared to the number of non-Chinese who were admitted: 99.1 percent of 2,581,820.[14] The Chinese were considered an inferior race without possibility of assimilation. As early as 1854, a California court ruled that Chinese could not testify against whites, reasoning that because Indians were prohibited by law from testifying and because the Chinese belonged to the same race, the law should apply to them as well.[15] Also in this period, the African American population grew by about 948,500, but immigration from Europe accounted for nearly three times the population growth of African Americans. Thus social Darwinian and Teutonist theory had combined to justify the disproportionate entry of northern European populations to the United States. The subsequent further development of eugenic theory and practice would combine to deepen the contradiction between the rhetoric and the reality of American democracy. The immigration restrictions of 1924 held in form until the Cellar Act of 1965, which finally allowed immigrants to be admitted to the United States by their order of application.

The Stage Is Set

We have now considered the pseudoscientific developments, beginning in the mid–nineteenth century, that prepared the way for international eugenics movements. Herbert Spencer provided an ideological underpinning for unregulated and brutal capitalism in the form of social Darwinism.

The Comte de Gobineau warned that the maintenance of the world's great empires was possible only by maintaining racial harmony within nations. Finally, Francis Galton argued that within a nation, genius and general ability were hereditary. He suggested forcefully that the state needed to take a role in fostering the growth of its best germplasm to prevent dysgenic decline. The ideas of these European scholars took root in minds of the American public, particularly among its young Anglo-Saxon/Teutonic elite.

These ideological tools appeared at a time when the United States was preparing to become a world economic and political power. It had settled the slavery question, although the South had reestablished the rule of the former slave masters over African Americans via Jim Crow. In addition, the resistance of the American Indian nations to colonization in the western states (let none of us forget Wounded Knee) had been destroyed. No one could doubt that the "manifest destiny" of the United States was strongly influenced by racist philosophy. By 1898, the United States would take its first step toward becoming a world power with its victory in the four-month war with Spain. This victory, like the victory against the Plains Indians, was made possible in part by the contributions of the African American soldiers of the 9th and 10th Cavalries, known as the Buffalo Soldiers, in Cuba. Ironically, the African American soldier would soon be relegated to a second-class role in the military by the same racist theories that had driven manifest destiny to begin with. Indeed, the beginning of the twentieth century would be a dark time for non-Anglo-Saxons in the United States.

PART 3

Applications and Misapplications of Darwinism

All science earns its right to existence in its application to real-world problems. In part 2 we discussed the birth of Darwinian science and saw how it was co-opted by racist ideology. This co-optation was possible, in part, because the theory had been born incomplete. As we have seen, the evolutionary world-view lacked a workable theory of heredity. Such a theory was developed subsequently over a period of approximately seventy years, during which the Mendelian rules of particulate inheritance would be reconciled with natural selection. The new disciplines of population, quantitative, and evolutionary genetics would combine to form what is now known as the neo-Darwinian synthesis.

The incompleteness of evolutionary theory did not, of course, prevent its being used to make predictions in a wide variety of fields. Scientific theories, by definition, are never complete; yet we use them to attempt to understand our world. This process need not be tragic, particularly if scientists apply the theories with the limitations in mind. Unfortunately, the Victorian eugenicists did not meet this test. The battle for the scientific high ground in eugenics was fought by some of the greatest minds in evolutionary biology. Villains and heroes emerged. By the 1930s, the misapplication of population genetics would be embraced wholeheartedly by some of the most powerful racist states the world has ever known. The result would be sterilization and extermination programs carried out under the flag of protecting the evolutionary future of the world's racial elite. If one wishes to see a powerful example of how seemingly esoteric scientific theory can impact the lives of millions, here is the case.

Mendelism, the Neo-Darwinian Synthesis, and the Growth of Eugenics

Early Developments of the Neo-Darwinian Synthesis and Racial Theory

As we have seen, the fatal flaw of all late-nineteenth-century arguments concerning the relationships between intelligence, human culture, and race was the lack of a correct model of the inheritance of genetic traits and how these traits related to evolution. Francis Galton's discussion of heredity and genius anticipated some of these basic problems. However, he lacked the sophistication required to fully grasp the true nature of complex phenotypes. In addition, he made his arguments in the context of the class, race, and gender biases operating during this period, vitiating their scientific value. The solution to this problem was the unification of genetics with Darwin's theory of evolution.

Darwin's understanding of heredity was embodied in his theory of pangenesis. He thought that "pangens" circulating in the bodies of the parents united during fertilization to produce a mixture of maternal and paternal traits in the offspring. This theory explained, for example, why a person's adult height was the average of that of both his parents and, in an example particularly relevant to racial theory, why the skin color of a mulatto was intermediate between that of the white and black parents. Darwin thought that all traits were determined in this additive way.

Mendel's laws of particulate inheritance had already been formulated before Darwin published *The Origin of Species*. Mendel had conducted experiments with pea plants in 1856. These plants were bred such that they always produced the same specific trait (true breeding). Mendel crossed different varieties of these true breeding plants to discover the laws of particulate inheritance. However, the report on these experiments did not appear until 1866, and even then Darwin did not read Mendel's seminal paper describing particulate inheritance. Even if he had read the paper, the incorporation of Mendelism into Darwinian thinking would not have been straightforward. In fact, its incorporation took about thirty years of work by some the best minds of the twentieth century. In addition, the amount of information available concerning inheritance did not guarantee successful integration.

Mendel's experiments demonstrated the existence of several genetic phenomena crucial to understanding the inheritance of biological variation. Of particular relevance to the idea of race were the concepts of biparental inheritance, dominant and recessive alleles, and independent assortment of genetic traits. After 1915, the American geneticist Thomas Hunt Morgan and his coworkers would demonstrate that independent assortment of traits occurred because the genes for those traits were on separate chromosomes of a cell. The process of meiosis, or cell division, in which gametes (reproductive cells) are formed both reduces the genetic complement by half and also recombines all genes found on separate chromosomes (or those located far apart on one chromosome). Genetic recombination occurs because meiosis shuffles the maternal and paternal chromosomes into new combinations. In addition, during the early phases of meiosis, pieces of chromosomes actually break off and exchange between maternal and paternal chromosomes. This phenomenon is called "crossing over," and it can bring together new genetic combinations not predicted from the original chromosome arrangement. Crossing over is more likely to occur between distant portions of chromosomes.

Adult human beings have two sets of 23 chromosomes, one set from each of the parents (the principle of biparental inheritance). Each parent thus contributes a gamete consisting of one set ($1N$, where N is the number of chromosomes) of these sorted chromosomes, which, when joined with the gamete contributed by the other parent, forms their offspring, who is now diploid ($2N$). In any diploid organism, the number of potential chromosome combinations is 2^N. The fruit fly, *Drosophila melanogaster,* has 3 chromosomes, so the number of possible combinations is 2^3, or 8. The 23 chromosomes of human beings give rise to 2^{23}, or 8,388,608, possible combinations.

In addition, each one of these chromosomes contains individual genes; and the position of a gene on its chromosome is called its locus. These loci are responsible for the production of the eventual phenotype, or physical traits, we find in the organism. Each locus harbors at least two different forms of a gene, called alleles. For example, in *Drosophila* one of the first alternative alleles discovered was the allele for white eyes. Fruit flies normally have brick-red eyes; however, the white-eyed mutant fails to produce the red pigment. Subsequent genetic analysis of the white-eyed phenotype would show that each locus may harbor several alleles; for example, there are at least eleven alleles that can form the white-eyed phenotype. Thus, if we consider the number of chromosomal combinations and the number of allelic variations in humans, we can see that even simple Mendelian principles can lead to complex inheritance of biological traits.

Mendel's laws of particulate inheritance explain the transmission of a surprisingly large number of human genetic traits (such as eye color, the ability to taste certain chemicals, widow's peak, color blindness, and some disease conditions such as hemophilia, dwarfism, or sickle cell disease). However, Mendel's laws do not readily explain more-complex traits, such as overall fitness, which is of central interest to evolutionary biologists. In addition, the laws do not readily explain behavioral traits, such as intelligence, of central importance to psychometricians (like Galton and Karl Pearson).

Ironically, Mendel's rules as he originally proposed them would not even have explained variation in skin pigmentation because pigmentation is determined by several genetic loci, which have equal and additive effects on the phenotype. Such characters are called quantitative traits. If Galton had been correct in apportioning human intellect into sixteen grades, which of course he was not, genetic rules would have predicted a minimum of six to seven genes impacting that phenotype.[1] Also, as the number of genes involved in producing a trait increases, so does the potential for gene-gene interactions (epistasis) and gene-environment interactions. These complications tend to smooth out the classes of possible phenotypes, such that we tend to observe continuous phenotypic distributions for such characters as height, weight, complexion, behavior, "intelligence," and so on, all quantitative traits. The existence of continuous phenotypic variation is why the Darwinians did not think that Mendel's laws accounted for the observed complexity of inheritance.

Several important discoveries and theoretical developments between 1900 and 1918 allow the Mendelians and Darwinians to begin building evolutionary genetics. Evolutionary genetics would have crucial importance for the new genetically based theories of race, in part, because some of the scientists involved in the basic evolutionary genetic research were also racist (for example, Francis Galton and the American eugenicist Charles B. Davenport). The concepts at issue had direct relevance to the question of the origin and maintenance of biological variation in species and thus to the race question. One of Davenport's students at the University of Chicago, William Castle, demonstrated that selection acting on continuous variation was effective in changing a character to a new stable level beyond the limits of pre-existing variation. This principle had been known intuitively for centuries; farmers had selectively bred animals and

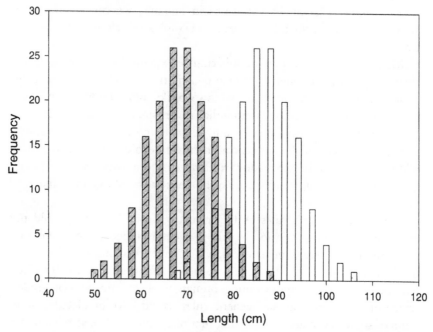

Figure 7.1. The result of directional selection on a hypothetical character: ▨▨▨▨▨ = original distribution, ☐☐☐ = distribution after directional selection. The figure shows an example of directional selection on a quantitative characteristic in a hypothetical population. Directional selection occurs when specific characteristics of only a portion of the population are selected for. The phenotypic means of population groups are important for understanding this response to selection: the mean phenotype of the entire population in the parental generation (which includes selected and nonselected individuals), the mean of the individuals selected to found the next generation, and the mean phenotype among the progeny of the selected parents. If the selection pressure continues in further generations, then the character distribution will continue to move to the right, hence the term directional selection.

agricultural varieties of plants by breeding beyond the limits of pre-existing variation. (Darwin, in *The Origin of Species*, had used animal breeding to explain the action of natural selection.) Castle formalized this principle, which would later be called directional selection (see figure 7.1).

Directional selection has been used explain how new skin color phenotypes, determined by at least six genes, developed during human migration. Human populations were originally almost certainly dark skinned (melanic), because they evolved in equatorial Africa. During the subsequent migration out of Africa, selection must have favored new skin colors beyond the limits found in the original human populations.

By 1918, the scientific consensus was that Mendel's principles could account for continuous variation. The subsequent full synthesis of Mendelism and Darwinism would occur mainly in the work of Sir Ronald A. Fisher, J.B.S. Haldane, and Sewall Wright. Again, the skin color example comes to mind. The application of Mendel's rules to multiple gene inheritance predicts that the offspring of parents with different skin tones should be intermediate in tone. Thomas H. Morgan and his coworkers would also discover the origin of new genetic variants in the form of mutations. They showed that Mendelian rules also applied to these very small variations. This finding was consistent with the Darwinian idea that evolution occurred gradually within a species. Thus, human variation might be the result of the accumulation of small genetic changes over relatively long periods of time.

Population Genetics

Population genetics, the study of the frequencies of genes in populations, is directly relevant to racial theory. The nineteenth-century naturalists studying race unknowingly were concerned with underlying genetic variation in the human species, although they lacked the formal rules to explain how that variation originated or was apportioned. Some of these rules are now known: for example, the Hardy-Weinberg equilibrium principle, the mathematical theory of natural selection, and the population genetics of inbreeding and genetic drift.

The Hardy-Weinberg equilibrium principle states that in the absence of evolutionary mechanisms acting on a population (for example, natural selection, mutation, nonrandom mating, migration), gene frequencies in that population will remain stable from one generation to the next. An analysis of the action of natural selection and genetic drift is impossible without this principle, which was derived independently by two men. Wilhelm Weinberg derived an expression in 1908 for the expected frequency of a genotype at equilibrium for a genetic locus with two alleles, and English mathematician Godfrey H. Hardy would independently apply this expression to the human blood-group alleles about six months later.

There are three main blood-group antigens (A, B, O). Antigens are proteins found on the red blood cell surface that stimulate an immune response. In a randomly mating population containing alleles for all three antigens, the only possible genotypes are AA, BB, AB, AO, BO, and OO; and we can calculate the expected frequencies of these genotypes.[2] The Hardy-Weinberg equilibrium principle predicts that in a large population, in the absence of external forces, the frequencies of these alleles will

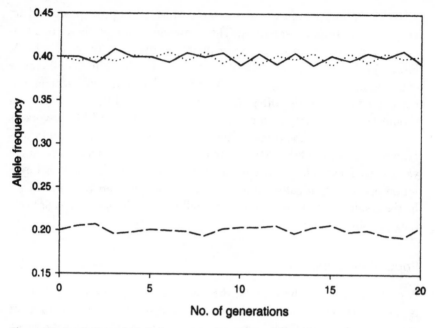

Figure 7.2. Predictions of allele frequency according to the Hardy-Weinberg equilibrium principle: —— and •••• = A and B alleles, — — = O allele. This figure shows that when the Hardy-Weinberg equilibrium is in effect, the frequencies of the A and B alleles and the O allele remain unchanged at 0.40 and 0.20, respectively.

remain constant through time (see figure 7.2). If the Hardy-Weinberg equilibrium were maintained over the entire human species, then we would see no genetic differentiation in our species and hence no differentiation remotely approaching the degree required for the formation of races.

There are at least two mechanisms that can disturb the Hardy-Weinberg equilibrium and rapidly change gene frequencies in populations: natural selection and genetic drift. The action of natural selection against a particular allele is relatively simple. Darwin's original definition of natural selection can be rephrased as the differential reproductive success of a genotype. Differences in reproductive success arise from differential survival or differential reproduction or both. Together, survival and reproduction determine a genotype's fitness. In the simplest case, individuals with a particular recessive, disfavored allele might be less successful in either survival or reproduction than individuals with the corresponding dominant, favored allele. The dominance or recessiveness of an allele is relative. The terms refer to whether the final phenotype will exhibit the given allele in the heterozygous case. For example, in a cross of a homozygous blue-eyed person and homozygous brown-eyed person, all offspring

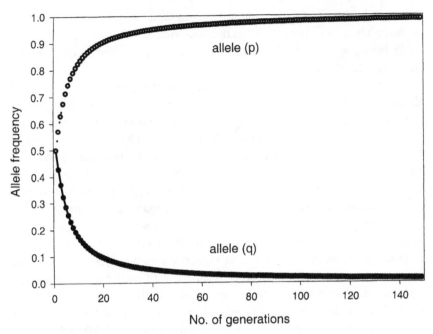

Figure 7.3. Allele frequency changes due to natural selection: ——•—— = allele p, ——○—— = allele q.

will be brown-eyed. This result indicates that brown eye color is dominant over blue eye color. Utilizing some algebra we can show that the population will become fixed for the favored gene and that the recessive allele will disappear.

Figure 7.3 compares the fates of two hypothetical alleles, one dominant (q) and one recessive (p), which start at equal frequency (p = q = 0.50). If the dominant allele performs 50 percent better than the recessive allele (which corresponds to a selection differential, s, of 0.50), the recessive allele is drastically reduced in frequency, although it persists for some time at low frequency. It is this type of selection scenario that Francis Galton was referring to in *Hereditary Genius* when he raised the specter of dysgenesis. Galton was able to show that when more-intelligent and abler individuals delay their marriage and reproduction as compared to the lower classes, natural selection would favor the lower classes. Note that Galton assumed that the social class of individuals was inherited genetically.

To illustrate this scenario, Galton gave the example of the decline of the Spanish empire due to the Inquisition. He claimed that over the three hundred years of suppression of free-thinking in Spain, more than 340,000 genetically "valuable" people were imprisoned, exiled, or executed. He concluded: "It is impossible that any nation could stand a policy like this, without paying a heavy penalty in the deterioration of its breed, as has notably been the result in the formation of the superstitious, unintelligent, Spanish race of present day."[3]

Galton's scenario lacked the sophistication of the formal mathematical theory of natural selection. If he and other eugenicists had understood this limitation, they would have seen that eliminating deleterious human genes by selective breeding is not so simple. In fact, in 1917 English geneticist R. C. Punnett published a paper demonstrating the fallacies of eugenics theory. He examined the elimination of feeblemindedness in humans, utilizing genetic selection theory to show that although the initial reduction in gene frequency is rapid, the trait persists for many generations.[4]

The difficulty of eliminating undesirable traits is made clearer when one examines the impact of the magnitude of the selection differential on the number of generations required to reduce a gene's frequency (this issue was examined briefly in chapter 6). In figure 7.3 a very high selection differential was used to reduce the frequency of the recessive allele (s = 0.50). Even under these conditions, reduction of the gene frequency from 0.50 to 0.10 took forty generations, and reduction from 0.10 to 0.05 took another ten generations, giving a total time for reduction to 0.05 of 1,500 years for humans. However, the selection figure of 0.50 would have been impossible to achieve for human civilizations. If the more realistic figure of 0.10 or less is used—that is, if the difference in reproductive success favoring the "intelligent" over the feebleminded is less than 10 percent—the reduction of feeblemindedness would take even longer. Moreover, we must remember that many of the early eugenicists felt that more than 90 percent of the population was feebleminded, in which case an even greater effort would be required to reduce the frequency of the gene. Few political programs in modern history have lasted 50 years, let alone 1,500. The idea that feeblemindedness could be selectively bred out of a population by these techniques is hopelessly flawed. This may explain why racist state machines that wanted to alter their countries' genetic composition attempted to do so not by increasing the breeding of the "fit" but rather by sterilizing or exterminating the undesirable, as in Nazi Germany.

By the mid-1920s, population geneticist Sewall Wright would propose that small population sizes could randomly alter gene frequencies and that this effect, called genetic drift, was an important factor in evolution. Genetic drift results from sampling error. The random sampling of gene frequencies from western Africa due to the Atlantic slave trade may serve as an illustration of genetic drift's effect on human racial composition. Only a small percentage of Africans were sold into slavery. In addition, mortality during the Middle Passage and "seasoning" in the New World reduced the population even more. Genes that were not directly associated with survival under these conditions would have been expected to randomly drift in frequency.

Sewall Wright's ideas about the action of random genetic drift in evolution were not fully appreciated until much later. Pre–World War II eugenics theory was dominated by the concept of natural selection. The theories of the classical school of population genetics, which held that the vast majority of mutations were deleterious, were particularly well suited to eugenics theory. The classical school expected that natural populations would show little genetic variability, because natural selection would be acting against the deleterious alleles and reducing them to low frequencies. As a consequence, populations were predicted to exhibit a typical phenotype called the wild type. The development of this type of population genetics theory contributed to the growth of eugenics. Many of the classical theorists, such as R. A. Fisher, were also eugenicists. To the classical theorist, racial variation was a case in point of the deleterious character of allelic variation. Thus, the eugenicists championed the idea that selective breeding principles should be used to direct the reproductive policies of entire nations. To eugenicists, the Nordic/Aryan races represented the wild type or the most "fit" components of the genetic variation within the human species. In the United States, the geneticist Charles B. Davenport would become the champion of this movement.

Charles Benedict Davenport and the Eugenics Record Office

In an October 10, 1910, letter to Francis Galton, Charles Benedict Davenport wrote: "There has been started here a Record Office in Eugenics; so you see that the seed sown by you is still sprouting in distant countries. . . . though our work is mainly in 'negative eugenics' we should put ourselves in a position to give positive advice. *We cannot urge all persons with a defect not to marry, for that would imply most people,* I imagine, but we hope to be able to say, 'despite your defect, you can have sound offspring if you will marry thus-and-so' [italics added]."[5]

From the inception of the American eugenics movement, Davenport was recognized as its "pope." This movement began six years after students of evolutionary theory rediscovered Mendel's work. Davenport had originally been a biometrician, having spent time in England as a graduate student learning the principles of eugenic breeding techniques under Galton and Karl Pearson. Davenport had been appointed to a prestigious faculty position at the University of Chicago in 1898, and he spent his academic year in Chicago and summered at Cold Spring Harbor Biological Laboratory in New York. As a professional biologist, Davenport was too aware of the general developments in biological theory to discount Mendelism. He became a close personal friend of geneticist Thomas Hunt Morgan. Clearly, he was exposed to the new developments in genetics. In fact, Davenport published several important basic research papers that contributed to the eventual unification of Mendelism and Darwinism.

In 1906 Davenport insisted that the American Breeders Association begin to study eugenics, and by 1910 Davenport had convinced the wife of railroad magnate Edward Henry Harriman to support the foundation of the Eugenics Record Office (ERO) at Cold Spring Harbor. The ERO's second in command was the fanatical Harry Hamilton Laughlin, an epileptic who refrained from having children because of his devoted belief in eugenics. The ERO always received between 13 and 29 percent of the total budget of the Station for Experimental Evolution (adjacent to the Biological Station). This meant that the ERO was well funded, particularly when compared with other scientific projects of this period.

The goals of the ERO were

- to become a repository and clearinghouse for data on human genetic traits,
- to build up an analytical index of traits in American families,
- to study the forces controlling the hereditary consequences of marriage-matings, differential fecundity, and survival migration,
- to investigate the manner of inheritance of specific human traits,
- to provide advice concerning the eugenic fitness of proposed marriages,
- to train fieldworkers to gather data of eugenic import,
- to encourage the creation of new centers for eugenics research and education, and finally
- to publish the results of research and to aid the dissemination of eugenic truths.

To collect the data necessary to study the inheritance of human genetic traits, the ERO trained a corps of fieldworkers who were to "diagnose" hereditary defects in people they interviewed. Each summer the ERO ran a course (with Davenport and Laughlin as chief instructors) for these workers. The topics included endocrinology, Mendelian genetics, Darwinian theory, elementary statistical methods, and eugenic legislation. Students also became familiar with the various mental ability tests (Binet, Yerkes-Bridges, Army Alpha and Beta) and learned how to administer and interpret them. Training also included the memorization of the categories of mental defectives, criminals, epileptics, and skin and hair colors and instruction on how to make anthropometrical measurements such as cranial capacity.

The fieldworkers took trips to nearby hospitals and institutions for mental defectives. This was practical training because they were to gather their real data in mental hospitals and insane asylums, as well as in private homes. By 1917 the ERO summer institute had trained 156 fieldworkers (131 women and 25 men, mostly from upper-middle-class backgrounds).

The data gathered by the fieldworkers were published in reports and widely disseminated to important political and commercial centers. The ERO published a monthly newsletter for public consumption entitled *Eugenical News*. Davenport utilized questionnaires sent to prominent public officials and business people to gather data on what made them successful. These questionnaires were also a funding tactic, since they were accompanied by forms that allowed these "genetically superior" individuals to donate to the ERO. Thus the data that Davenport gathered on those of "superior" and "inferior" blood differed greatly in the manner of their collection. The superior sent in their own responses (and only superior people were contacted and informed of their superiority by means of the questionnaires), whereas the inferior were examined by the "trained" eugenics fieldworkers.

To facilitate the study of human traits, the ERO kept and indexed all incoming data in a "trait book." Many conditions felt to be human genetic traits were indexed: intelligence, ability at chess, feeblemindedness, insanity, manic depression, liveliness, moribundity, lack of foresight, rebelliousness, trustworthiness, irritability, popularity, rowdyism, and moral imbecility. The early eugenicists felt that all human traits were simple unit characters. The trait book was the means of studying the inheritance of specific human traits; it was used, for example, to determine whether a trait was autosomal or sex-linked, dominant or recessive. The ERO researchers made no attempt to distinguish between innate and learned capacities.

The focus on the study of fecundity and its relation to specific genetic traits is the core of eugenics research and its social political program. Eugenics is also premised on the idea that unworthy traits exhibit a positive genetic correlation with high fecundity. This connection has always been only an assumption: the genetic determination of complex behavioral traits is often unknown. The ERO, despite its rhetoric to the contrary, was always measuring phenotypic traits. Its primary techniques to establish the genetic character of a trait were to utilize the data gathered by its field-

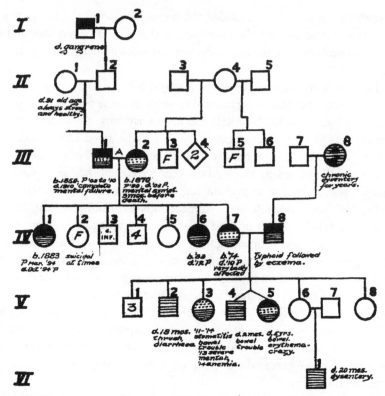

Figure 7.4. Sample pedigree from the Eugenics Record Office. Squares indicate males and circles, females; solid color, lines, or dots indicate pellagra symptoms. Davenport reasoned that pedigrees showing pellagra's appearance in successive generations were evidence that the disease was primarily genetic. From C. B. Davenport, "The Hereditary Factor in Pellagra," *Archives of Internal Medicine* 18 (1916): 18.

workers and to draw pedigree charts (see figure 7.4), which were used to infer genetic inheritance.

To be useful for inferring inheritance, such charts must be large enough to reliably calculate the ratios predicted by Mendelian inheritance mechanisms. However, even when sufficiently large, human pedigrees do not conclusively demonstrate genetic causality, because groups of related individuals often live in similar environments. The separation of genetic from environmental effects requires some form of experimental intervention, but the means to experimentally examine the genetic nature of phenotypic correlation are generally precluded from human studies. Davenport's colleague Thomas H. Morgan was able to establish the genetic character of his *Drosophila* traits because he could control both the mating of the flies and their environment. As much as the ERO might have wished for this ability, they never had it.

The ERO expected that its collection of data on human traits would allow it to advise individuals on the fitness of their marriages. In addition to advocating positive eugenics, the ERO wished to play a major role in protecting the nation from the untrammeled reproduction of the genetically unfit, and one method of accomplishing this was the passage of sterilization laws and immigration restrictions. Harry Laughlin authored the model eugenic sterilization law in 1924, the same year, not too coincidentally, that the Immigration Restriction League (IRL) succeeded in getting immigration restrictions passed. The model sterilization law was distributed to governors, legislators, newspaper and magazine editors, clergy, and teachers. The goal of the model was to assist states in enacting compulsory sterilization laws for supposed genetically defective individuals. Laughlin's far-reaching model law recommended sterilization for the following classes of individuals:

> The socially inadequate classes, regardless of etiology or prognosis [that is, regardless of the cause of the condition or the possibility of eliminating it by means other than gelding], are the following: (1) Feeble-minded; (2) Insane (including the psychopathic); (3) Criminalistic (including the delinquent and wayward); (4) Epileptic; (5) Inebriate (including drug-habitués); (6) Diseased (including the tuberculous, the syphilitic, the leprous, and others with chronic, infectious, and legally segregable diseases); (7) Blind (including those with seriously impaired vision); (8) Deaf (including those with seriously impaired hearing); (9) Deformed (including the crippled); and (10) Dependent (including orphans, ne'er-do-wells, the homeless, tramps, and paupers).[6]

A quick examination of these categories shows that only a small portion of the "defects" among these classes could be directly traced to a genetic condition. The fact that thirty states adopted sterilization laws for the "mentally defective" can be laid directly at the feet of Laughlin and Davenport of the ERO. In the United States, sixty-five thousand people had been sterilized against their will by 1968. Most of these individuals had been labeled as "mentally retarded" on the basis of IQ tests. The 1925 case of Carrie Buck, ordered sterilized under the newly passed Virginia law, has been well discussed.[7] Carrie Buck was an involuntary resident of the State Colony for Epileptics and Feeble-Minded and was to be the first person sterilized under the new law. A local group of Christians challenged the state's general right to mutilate a ward of the state. The ERO sent one of its fieldworkers to investigate the case first and then sent Harry Laughlin to present his "scientific" findings before the Circuit Court of

Amherst County. In his testimony Laughlin admitted that feebleminded-ness might be influenced by both environmental and genetic factors, but he suggested that Buck's feeblemindedness was clearly a result of the latter. The ERO testimony played a major role in the Supreme Court's eventual decision to order Buck sterilized. The Nazis would base their initial steril-ization efforts on these same American laws and would sterilize invol-untarily 2 million people (and would execute millions more in the death camps).

For the most part, the American eugenics movement was actually di-rected at whites like Carrie Buck rather than at blacks. Two works warned white Americans in this period of the desperate need to use eugenic mea-sures to improve the nation: Madison Grant's *The Passing of the Great Race* (1916) and Lothrop Stoddard's *The Rising Tide of Color against White Supremacy* (1920). Grant and Stoddard argued that the immigra-tion of the defective races of the world threatened to erode the United States' moral and intellectual character. This erosion would result from the greater reproduction of the genetically defective and from the mongreliza-tion caused by interbreeding between whites and these undesirables. Their solution was twofold: first, the use of eugenic measures to improve the Anglo-Saxon/Nordic component of the white race and, second, to close the floodgates to immigration by inferior races. Grant made it no secret that he felt the inferior races would be doomed once the United States adopted mandatory sterilization of the socially worthless. Grant and Stod-dard pointed to the so-called poor white trash, typified by the southern white, as exemplifying the degeneration of the Nordics. In arguing the Carrie Buck case, Laughlin and the ERO characterized this population as a "shiftless, ignorant, and worthless class of anti-social whites of the south."[8] The ERO would also attempt to use them as its test case for prov-ing that laziness was a heredity character.

Eugenics and Medical Fraud

The ERO's ideological positions resulted in its ascribing genetic causes to any negative attribute of its target populations. However, as the ERO reached its maximum influence, public health research began to erode its authority. For example, U.S. Public Health Service parasitologist Joseph Stiles would (with Rockefeller Commission support) demonstrate the true cause of southern "laziness." We have already seen that southern popula-tions exhibited high frequencies of infection with the hookworm *Necator americanus* (see chapter 5). Adults of the genus *Necator* burrow into the intestinal mucosa, where they consume blood at tremendous rates. The worms reproduce rapidly, producing thousands of eggs daily. The impact of the eggs is nontrivial, for they can overtax the immune response. *Necator* infection leaves patients anemic, lackadaisical, mentally deficient, and prone to opportunistic infection. Stiles found that the hookworm disease was associated with two types of environmental variables, poverty (the inability to afford decent shoes and sanitation) and sandy soil (versus clay soils). *Popular Science Monthly* for February 1903 reported the following:

> An important point claimed in these [Stiles's] investigations is that hookworm disease is especially prevalent among children, and that it not only interferes with their school attendance, but the children who are afflicted with the malady and who have gone from sandy districts to a city have the reputation among their teachers of being more or less backward and even stupid in their studies. All this agrees with well-established symptoms of the disease, for it is thoroughly established, not only by Dr. Stiles's investigations, but by observations in Europe and Africa, that hookworm disease stunts both the physical and the mental development.[9]

Despite this evidence, the ERO continued to use susceptibility to disease as evidence of genetic inferiority. In this regard, Davenport and the ERO would engage in one of the largest medical frauds of the twentieth century: the pellagra cover-up. Joseph Goldberger, a dedicated but poorly paid researcher for the U.S. Public Health Service, demonstrated in 1915 that pellagra resulted from vitamin deficiency. Goldberger's first crucial observation in his conquest of pellagra was the discovery that wherever the disease was endemic, it was never associated with well-to-do families. This observation eliminated the possibility that pellagra was a contagious disease. In addition he was able to show a correlation between the amount of dietary protein and the incidence and severity of the disease: the disease was more frequent and more severe in populations with low-protein diets.[10] Utilizing these observations, Goldberger was able to produce the disease in convicts by giving them unbalanced diets and to cure it by restoring protein and green vegetables to their diet.

The ERO, however, claimed that pellagra was hereditary. To foster this claim, Davenport and Elizabeth Muncey, a physician paid by the ERO, published two papers in 1913 in the *Archives of Internal Medicine* purporting to show a correlation between pellagra and other so-called genetic deficiencies such as insanity, mental retardation, imbecility, pauperism, and low IQ-test scores. In 1917, articles were included in *Pellagra III*, the final report of the Pellagra Commission of the New York Post Graduate

Medical School and Hospital, with which the ERO was involved. The report included tables of family pedigrees (prepared by Muncey) reputedly proving the hereditary character of pellagra. These pedigrees included other "genetic" diseases, such as insanity, hookworm, and suicide. The report was crucial in supporting those who wished to ignore the Goldberger results. Davenport was still considered one of the United States' greatest scientists. His opinion was essential backing for those determined not to make an effort to cure the pellagra epidemic. Despite Goldberger's eloquent experiments, scientific debate concerning the nature of pellagra continued into the 1930s. The intergenerational transference of negative physical and social conditions often gave the appearance to a naive geneticist that the external phenotype (laziness, mental dullness) was genetically inherited.

All parents initially give two things to the children that they raise, their genetic material and their lot in society (with all of its environmental detriments or benefits). The "geneticists" of the ERO never seemed to grasp this fact, even though ample population and quantitative genetic theory and data existed to make the point abundantly clear. Thus, from 1916 to 1941 more than seventy-five thousand Americans would die from pellagra (during the period in which records were kept by "race," 55 percent were nonwhite). In contrast, only fifteen people would die from this disease in 1968.[11]

Davenport and Race-Crossing

None of Davenport's falsifications made his biases more clear than his study of "race-crossing." Davenport assumed that race mixing would lead to "disharmonious impacts on the hybrids."[12] The phenotypic traits of the hybrids would not match the average of those of the individual races. This hypothesis was meant to rebut environmental explanations of deviation in the phenotype. Davenport may have been influenced by the work of Herbert Spencer Jennings at Harvard in 1908. Jennings (who himself later became a noted eugenicist) examined the impact of breeding pure lines of the protozoan *Paramecium* on their average size. Jennings isolated single individuals of average size within a strain and then bred the descendants under constant conditions for one hundred generations. He found that average size was unique to each lineage, although variation within each line was present. Furthermore, he attempted artificial selection on the pure lines and found that the average size of the lines did not change. This led him to conclude that the average size was strictly hereditary throughout the pure line. Today we know that the pure lines could not have responded to selection because by definition they contained no genetic variation (and his population size was too small to introduce a high enough frequency of mutations to create new variation). Jennings also performed no experiments in which he crossed the different pure genetic lines (which, contrary to Davenport's expectations, would have revealed overyielding, or a general improvement in the stocks).

An examination of Davenport's positions from the point of view of the prevailing ideology of the neo-Darwinian synthesis reveals them as thoroughly reactionary. Davenport's hypothesis relied on essentialist conceptions of race (for example, the idea that the average represents the harmony of the racial line) and nullified the predictions of the biometricians concerning continuous variation. It also revealed an ever present logical inconsistency in the thinking of eugenicists and psychometricians: the assumption that the results of highly simplified experiments (such as those conducted on protozoa) directly map complicated systems like humans. To achieve his results, Jennings needed to carefully maintain constant and equalized environmental conditions for his strains. No such conditions have ever existed for the human "races." Indeed, Davenport's collection and interpretation of his data in the race-crossing study were severely criticized in his own time. For example, Davenport was unable to verify any of his own hypotheses: "The Blacks seem to do better in simple arithmetic and with numerical series than whites. They also better follow complicated directions for doing things. It seems a plausible hypothesis, for which there is considerable support that the more complicated the brain, the more numerous its 'association' fibers, the less satisfactorily it performs the simple numerical problems which a calculating machine does so quickly and accurately."[13]

One can ask whether any evidence could have convinced Davenport that blacks were not inferior. He even realized that one could not use skin color as an accurate description of the genetic ancestry of a person, saying that it was impossible "from observation of skin color alone [to] draw an accurate conclusion as to the genetic constitution of a person."[14]

What this means is that Davenport must have understood that individual genetic traits are inherited independently. His own published works and his association with Thomas Hunt Morgan would have made the lack of such a conceptual connection impossible. However, as it does for many Euro-Americans, the subject of race tended to create a blind spot that objective scientific reasoning could not penetrate. The examples above reveal both his egregious scientific errors and his deceptions to save his political agenda. Davenport reduced to simple Mendelian factors complex genetic and environmental causes of dire problems. This he accomplished with full knowledge of the state-of-the-art genetic information at the time.

This body of theory was sufficient to allow more-objective scientists, such as Morgan, to realize that eugenics was pseudoscience moved by a political agenda. The motivation of Davenport and his coworkers to advance eugenics is clear. More puzzling were the motivations of Morgan and others who remained mostly silent while atrocities were perpetrated in the name of genetics. Personal loyalty to friends or funding from private endowments that advanced a eugenic agenda might explain some of this. Or perhaps the prevailing racist and classist ethos of the time forestalled their caring enough to take action. No one can think that Davenport's ERO could have kept up its operation if enough prominent scientists had vigorously opposed its efforts. Certainly, Morgan's reputation alone would have been enough to shake the eugenics movement at the core had he decided to denounce its methods and goals. The fact that many leading scientists were eugenicists and racists might also have prevented less luminary scientists from taking action. For example, the Galton Society, devoted to eugenics research in the United States, included many prominent scientists.[15] It is also important to remember that the American scholarly community was highly concentrated in the 1920s, in a way that it is not today. The greatest minds in disparate fields were all in a few intellectual centers, such as Boston, Chicago, and New York. New York, for example, attracted several active anthropological institutions with close contacts to major philanthropic sources of funding. The Galton Society and Cold Spring Harbor were the centers for the eugenicists and racists, and the New School for Social Research provided a liberal outlet for the faculty at Columbia. Both the American Museum of Natural History and Columbia University were at the core of anthropological activity and employed anthropologists of both liberal and conservative commitments.

As in American academia today, individuals representing the ideological diversity within the discipline were in close contact. But today the size and distribution of the scientific community have changed. The number of university and private positions for scientists is greater, and more federal research support, as opposed to support from private foundations, is available. The diversity of the modern research community makes it possible for one to take a contentious position without immediately destroying one's career. Finally, the desegregation of American universities has allowed hitherto marginalized minority populations to gain entry into the scientific research community. Their presence, while still small, has nonetheless forced a reexamination of the terms of the race debate in anthropology, genetics, and the social sciences.

The Opposition to Scientific Racism in Academia

The most serious opposition to scientific racism and eugenics in the early twentieth century came from the anthropologist Franz Boas and his students, including Alfred Kroeber, Robert Lowie, Alexander Goldenweiser, Paul Radin, Melville Herskovits, Ruth Benedict, Margaret Mead, and Otto Klineberg. Boas was of German Jewish ancestry and had originally been trained in physics. He received his training in physical anthropology under German pathologist Rudolf Virchow. Boas and his research group would shatter one of the central tenets of racist typology of human forms with their study of the head forms of immigrants and their offspring. This 1908 study examined the head forms of nearly eighteen thousand European immigrants and their foreign- and American-born children. The study was conducted for the U.S. Immigration Commission, which fully thought the results would support its limitation of immigration from southern and eastern Europe. The study was published as a report entitled *Changes in the Bodily Form of Immigrants* (61st Cong., 1908, S. Doc. 208), and it was released by Columbia University Press in 1912.

The report found that the American environment had a profound impact on the head forms of the children of immigrants; significant differences in cranial measurements were found between European parents and their American children. In addition, there were significant differences between the head forms of the children born in Europe and those born in the United States, and the differences were directly related to the amount of time the children had lived in the United States. Boas believed that the significant environmental inputs responsible for these differences consisted at least of nutrition and pathologies differing in Europe and the United States. Boas and his coworkers demonstrated in this study that the heritability of head forms must have been low. Boas, representing the minority view, raised questions of fact and methodology that could not be ignored by the leading racists of the period. In fact, it is likely that the American Galton Society was formed to counter the growing influence of the Boasians in anthropology.

The Popularity of Racist Ideology in the General Public

The period from the end of World War I to the start of the Great Depression was filled with social turmoil. Race relations in the United States were in a state of open warfare. African Americans who had fought in Europe during the war now had higher expectations for equal protection under the law and for the respect that was their due as human beings. The philosophy of Marcus Garvey and the mass popular movement organized as the Universal Negro Improvement Association (UNIA) had also stimulated greater pride for many African Americans in their African heritage ("Up you mighty race" and "Africa for the Africans"). In New York, the Harlem Renaissance (1922–1929), which featured great achievements for African Americans in music, art, and literature, was well under way. In addition, at least half a million African Americans had left the South for northern cities.

Euro-Americans were frightened by the political and social changes occurring all around them. For example, the period from July 13 to October 1, 1918, saw twenty-five major race riots across American cities, leaving one hundred dead and one thousand wounded. There were eighty-three recorded public lynchings that year. In 1923 the entire town of Rosewood, Florida, was destroyed by white racists. The United States was also in the grip of the Red scare. The American Communist Party had been formed, and labor unrest was growing across the nation.

Eugenics offered the middle-class white American a simple explanation for all this turmoil. From 1890 to 1924 over four thousand publications appeared on the subject of eugenics, of which sixteen hundred were directed at the lay public. In addition, a survey of 499 colleges and universities showed that, by 1929, 343 of them offered courses in eugenics. We already know that the eugenicists felt that all components of the human personality were preordained by genetic inheritance and, thus, that all of society's problems were the result of the prolific breeding of genetic inferiors. What most illustrates the appeal of this idea is the way it caught on among diverse elements of American society. The public appeal of eugenics should not, however, be confused with support in the community of professional science. At this point, the exact mechanism by which genetic features of populations were controlled had not been worked out. As with many scientific hypotheses whose applications conveniently fit into the social agenda of the rich and powerful, eugenics found its way into the lay discourse. People of all varieties preached its gospel: plant breeder Luther Burbank, popular champion of eugenics in California; Stanford University president David Starr Jordan; inventor Alexander Graham Bell; industrialist John Kellogg; dancer Isadora Duncan; and playwright George Bernard Shaw.[16] The eugenics societies even featured contests to choose well-bred babies.

It should be made clear that political ideology did not guarantee acceptance or rejection of eugenics. Herman J. Muller, for example, was a self-professed Marxist who originally felt that eugenics could produce the type of human being suitable for the dictatorship of the proletariat. Later,

Muller would describe eugenics as "lending a false appearance of scientific basis to advocates of race and class prejudice, defenders of vested interests of church and state, Fascists, Hitlerites, and reactionaries generally."[17] The entire ethos of the French and the early German eugenics movements was motivated by commitment to socialist principles. British geneticist J.B.S. Haldane would write a thorough critique of eugenics in 1938 from a socialist perspective. He pointed out that some "defects" had a simple genetic basis but that the human population was too complex for the elimination of such traits in a few generations.

Decline of the Eugenics Record Office

By 1940, progress in the study of genetics would seriously discredit both the ERO and the theories of eugenics. The ERO's records were eventually turned over to the Dight Institute for Human Genetics at the University of Minnesota. Despite Davenport's emphasis on rigorous scientific collection of data by fieldworkers, most of the information they collected was highly subjective and based on hearsay. Sheldon Reed, former director of the Dight Institute, would say that most of the data collected by the ERO were worthless from the standpoint of human genetics. In addition, the political activities of some of the ERO's members (particularly Laughlin's ties to Nazi officials) had drawn it into political disfavor. At this point the Nazis had already invaded Poland, and the world was beginning to witness the practical applications of eugenics firsthand. The combination of these two factors would lead the Carnegie Institution to cut off the ERO's funding.

CHAPTER 8

Eugenics, Race, and Fascism
The Road to Auschwitz Went through Cold Spring Harbor

The tragedy of Nazi Germany stands as the clearest example of what can happen if eugenics, racial hierarchy, and social Darwinism are taken to their logical conclusions. We have seen that by the 1930s American and British scientists differed among themselves in their evaluation of the legitimacy of eugenics and in the ulterior motives behind their allegiance to eugenics. Some, such as Sir Francis Galton, Charles B. Davenport, Harry H. Laughlin, Earnest Hooton, and R. Ruggles Gates, were clearly racist elitists. These individuals felt that not only the degraded darker human races but also 80 percent of "Caucasians" should be prevented from reproducing. By contrast, eugenicists like Karl Pearson, Julian Huxley, J.B.S Haldane, and Herman J. Muller, who were socialists, supported many of the same policies, but they did so with the idea of making the human species better able to help usher in the age of socialism. Despite the appeal of eugenics to wide sectors of the upper and middle classes, American democratic traditions limited its application in the United States. After all, its main aim was the eugenic improvement of the white race.

The Nazi regime, however, would apply the theoretical prescriptions of the eugenicists on a scale never before seen in history. The goal of the Nazi eugenics movement, known as the race hygiene (*Rassenhygiene*) movement, was to remove all inferior genetic material from the German people. The groups seen as inferior included Africans, Jews, Gypsies, Slavs, homosexuals, the feebleminded, paupers, communists, "criminals," and so on. The Nazi control of the state political machinery and the support of large sectors of the German ruling and middle classes allowed the Nazis to sterilize and murder millions in implementing this program. Although the German race biologists borrowed ideas heavily from England and the United States, they also developed their own unique contributions to racist theory. These were made possible by three aspects of German history: first, Germany lacked a history of support for liberal democratic social principles; second, Germany had suffered a disastrous defeat by the Allied Powers in World War I; and third, the form of social Darwinism that took root in Germany was particularly virulent. Social Darwinism was so virulent in Germany because the German understanding of evolution was derived more from the thinking of Ernst Haeckel than that of Charles Darwin.

Darwinism Comes to Germany

Ernst Heinrich Haeckel (1834–1919), the son of Prussian civil servants, developed a strong nationalist sentiment early in life. Bismarck had yet to unify the German confederation when Haeckel read Darwin's *Origin of Species* in 1860. As a young professor, Haeckel immediately began converting large numbers of students to Darwinism. In fact, Haeckel would become Germany's leading naturalist and a scientist of great international reputation. To Haeckel, Darwinism encompassed a wide range of disciplines. He immediately applied it to society, seeing struggle and selection as pushing people irresistibly to higher cultural stages. Haeckel formed a universal law of cultural development based on Darwinism. Through this law, he believed, would come about the evolution and unification of a Teutonic Germany. He also incorporated German "volkism" into his view of human evolution. German *völkisch* ideology of the nineteenth century combined the concept of the racial unity of all Teutonic people with the idea of a higher "essence." This "essence" was the source not only of all creativity, depth of feeling, and individuality but also of unity within the German people. Thus, Haeckel's idea of evolution was not really akin to Darwin's; whereas Darwin emphasized natural selection and the impersonal action of nature, Haeckel associated evolution with a teleological and mythological progressivism.

Haeckel was an Aryanist, and he separated the human species into definite races representing stages in a progressive evolution leading to the highest form of human, the Aryan. Haeckel explained the separateness of the races by utilizing his biogenetic dictum: "ontogeny recapitulates phylogeny." Haeckel proposed that the ontogenetic history of the individual must repeat the evolutionary history of its ancestors. He saw the ontogenetic history of the Aryan race in the lower, more primitive races. To Haeckel, the differences between the human races were greater than the average difference between humans and other primates. The higher races were destined to eliminate the lower races in the struggle for survival. Haeckel was strongly anti-Semitic. He felt that Jews had inborn genetic characteristics that were resistant to change. He even joined Houston Stewart Chamberlain in claiming that Christ's merits resulted from the fact that he was only half Jewish.[1] Haeckel felt that the historical persistence of anti-Semitism must have a rational basis: he could not believe that the highly evolved moral sense of the Aryan race was capable of maintaining an irrational prejudice. Thus, he surmised that anti-Semitism must have resulted from the actions of the Jews themselves. In his public discourse he demanded that the Jews assimilate into German culture, although he strongly felt that eastern European Jews could not be assimilated. Although Haeckel never called for the murder of German Jews, one can readily see that his views were consistent with the views of those who eventually did.

Haeckel's evolutionary views were supported by the work of the Austrian-born geneticist and embryologist August Weismann. Weismann's seminal contribution to genetics was the demonstration in 1883 of the continuity of the "germplasm" (genetic material). He was able to show that the germplasm is unaltered by environmental changes that affect the somatic tissue. Thus, mice that are born with tails but have their tails cut off in adulthood still give birth to mice with tails. Therefore the adult animal must be passing on something that is unaltered by environment, the germplasm. This result countered the persistent Lamarckianism of the Europeans, which proposed that acquired (or environmental) influences on the germplasm might be inherited. The importance of the continuity of the germplasm to eugenics theory is clear. If environmental influences cannot alter inborn traits, and if social and cultural traits are inborn, then the elimination of alleles for deleterious social and cultural traits requires a program of selective breeding. In other words, if the cultural traits of the Jews were inborn, then no modification of culture or environment would be capable of Aryanizing the Jews. In addition, if an undesirable race were out-reproducing the Aryan race, then means would have to be taken to control or retard the proliferation of that race. Haeckel had a major influence on the German eugenics movement, which in turn directed the racial thinking of Adolf Hitler and other Nazi race theorists.[2]

The German Race Hygiene Movement

Alfred Ploetz was the leader of the first race hygiene movement in Germany (he coined the term in 1895), and the movement's concerns were much like those of the international eugenics movement. To Ploetz, races were groups of interbreeding populations that over the course of generations had achieved similar physical and mental characteristics. Ploetz included in his thinking about race not only genetic composition but also optimal population size. His ideas paralleled the thinking of the American population biologist Raymond Pearl, who later became an important eugenicist. Ploetz arrived at the need for controlling population growth in general precisely for the purpose of slowing the increase of genetically inferior stocks. He saw the Aryan race, in particular the Germanic stock of it, as representing the cultural race par excellence. Ploetz was not entirely anti-Semitic. He did not see Jews as a biologically distinct race; he felt instead that, in Germany, they were overwhelmingly Aryan in racial composition.

Ploetz would play a major role in founding the organized eugenics movement, which in Germany in the period between 1904 and 1918 was called the Wilhelmine race hygiene movement. Its goals were similar to those seen throughout the world (prevent the breeding of inferior social traits, support the breeding of superior traits), and its composition was mainly professional and multiethnic (it still contained many Jewish members). It was not initially attracted to American-type sterilization legislation. In 1904, Ploetz founded the eugenics-inspired journal *Archiv für Rassen- und Gesellschafts-biologie*. On the editorial board were future Nazi race scientists Eugen Fischer and Fritz Lenz. The first issue of the *Archiv* was devoted to Haeckel and Weismann, and Haeckel was prominently cited throughout.

World War I would profoundly influence the thinking of the Wilhelmine movement. In particular, there was much concern about the growing threat of the perceived Slavicization of Europe. The growing "yellow peril" from China and Japan alarmed Munich physician and ardent eugenicist William Schallmayer. Schallmayer (also profoundly influenced by Haeckel) warned Germany that nations had declined in the past because they did not know how to avoid biological decay (in this we can see the influence of the Comte de Gobineau). Schallmayer felt that the nation had to protect its germplasm, and he was particularly concerned with the growing practice of birth control among upper-class women. Through birth control, the hereditarily fit were limiting their families while the genetic pauper classes within Germany were increasing at an unchecked rate. This problem was magnified for Schallmayer by the increasing Slav population and the "yellow" menace. The movement supported a variety of programs designed to provide incentives for increasing the family size of the favored classes.

The defeat of Germany in World War I slightly changed the emphasis of the race hygiene movement. After the war, the eugenicists felt they needed to implement programs that would prevent the decline of the German racial stock in the wake of the continuing assault of its enemies, particularly western Europeans and Russians. In the period following the war, the Munich branch of the race hygiene movement would develop under the ideological leadership of the future Nazi Fritz Lenz. Lenz was even more enthusiastic about the superiority of the Aryan race than Ploetz was. His enthusiasm may have resulted from his study with the anthropologist and later ardent Nazi Eugen Fischer. Lenz viewed races in a strictly hierarchical sense, placing Negroids at the bottom, Jews somewhere in the middle, and Nordics at the top. He was favorably impressed by Madison Grant's *The Passing of the Great Race*. Lenz considered liberal politics, money-making, and sexual proclivities as racial and genetic characteristics of the Jews. Lenz was a political conservative, belonging to the far right German National Party after 1918, and he opposed all democratization of the Weimar Republic.

The Munich branch of the race hygiene movement is significant in that much of the strength of the early Nazi movement came from the Bavarian region rather than from the more industrialized areas, such as Prussia. Berlin had a stronger socialist labor union presence. As economic depression and crisis deepened in Germany, the Munich wing of the race hygiene movement gained favor, particularly for its view that eugenic sterilization was a means to aid the economic crisis. In the Weimar Republic, sterilization measures became popular, for the state simply could not afford to pay for the upkeep of "mental defectives." The German scientific and medical community paid careful attention to the implementation of sterilization legislation in the United States. In the 1920s, Germans felt that the alternatives, institutionalization and work colonies, would be a more efficacious way to prevent the reproduction of the unfit. However, in 1932, the Prussian Health Council drafted a voluntary sterilization law that required proof that the defective trait in question was, in fact, genetic. Ironically, geneticist Richard Goldschmidt, then director of the Kaiser Wilhelm Institut für Anthropologie und Eugenik, was a supporter of the 1932 law. Goldschmidt, of Jewish ancestry, would later leave Germany for the United States.

Eugen Fischer was also a member of the Munich movement. He had received his medical degree at Freiburg in 1898, where he developed an interest in anthropology. He became famous for his work on the inferiority of "negro-white" hybrids in Africa. Anti-African racism had been fueled in Germany by German colonialism. For example, in 1908 in the German colony of South-West Africa (now Namibia), all existing German-African marriages were annulled and such marriages forbidden in the future. The Germans involved were deprived of their civil rights. Fischer, a docent in anatomy at the University of Freiburg, began to investigate the "bastards" of Rehoboth in German South-West Africa. These were persons of "mixed blood," born mainly of unions between Dutch (Boer) men and Hottentot women. In his book *Die Rehobother Bastards und das Bastardsiseirungs-problem beim Menschen* (The bastards of Rehoboth and the problem of miscegenation in man, 1913), Fischer writes of these people: "We should provide them with the minimum amount of protection which they require, for survival as a race inferior to ourselves, and we should do this only as long as they are useful to us. After this, free competition should prevail and, in my opinion, this will lead to their decline and destruction."[3]

Die Rehobother became a classic for white supremacists throughout the world. In it Fischer described Negroes as being less intelligent, creative, and spiritual than whites. He prescribed (as had Galton before him) a program of just but stern treatment for them. In 1927 Fischer became director of the newly founded Kaiser Wilhelm Institut. He soon became the foremost theorist of race hygiene in Germany and one of the most recognized eugenicists in the world. In 1929, Charles B. Davenport asked Fischer to become chair of the committee on "racial crosses" of the International Federation of Eugenics Organizations. In 1932, Davenport further showed his confidence in Fischer by nominating Fischer to succeed him as president of the federation. Fischer declined, because of other commitments, and Dr. Ernst Rudin (of the Munich race hygiene school) was elected. In 1937, Eugen Fischer had the opportunity to act on his anti-African racism within Germany when he played a major role in the illegal sterilization of all German "colored" children, the offspring of colonial African and African American troops of the French occupation force (1918–1930). The sterilizations of these children occurred before the reich officially passed its involuntary sterilization law and were never officially recorded.

The race hygiene movement, particularly the Munich section, profoundly influenced Nazi racial theory. In 1923 Adolf Hitler read the second edition of the textbook by Erwin Baur, Fischer, and Lenz titled *Menschliche Erblichkeitslehre und Rassenhygiene* (The principles of human heredity and race hygiene) while he was imprisoned in Landsberg. Hitler incorporated these racial ideas into *Mein Kampf*. The Nazis quickly put these principles to work. In December of 1931, Heinrich Himmler ordered that all members of the Storm Troopers (SS) had to obtain permission from the newly constituted SS Race Bureau to marry. Under this order, the bureau would grant or refuse permission solely on the basis of criteria of race and hereditary health. Professor Lenz, who had obtained a university chair on race hygiene in Munich, commented that this was a "worthwhile exercise."[4]

American Scientists and Hitler

"To that great leader, Adolf Hitler!" said Clarence Campbell, an American representative at the reception for the 1935 International Population Congress in Berlin.[5] In the February 27, 1940, issue of the *Richmond Times-Dispatch,* Joseph S. DeJarnette, a member of the Virginia Sterilization Movement wrote, "The Germans are beating us at our own game."[6] Comments such as these would indicate that eugenics and fascism were from the onset international political and ideological movements. For example, Davenport, Laughlin, Grant, Pearl, Frederick Osborn, and the rest of the United States' scientific racists had contributed both ideologically and materially to the rise of the Nazis. To the fascists, Judaism was not a culture or religion but a set of Mendelian traits. Otto Wagener, head of the Nazi Economic Policy Office from 1931 to 1933, wrote that Hitler was favorably impressed by American eugenics. Hitler said: "I have studied with great interest the laws of several American states concerning prevention of reproduction of people whose progeny would, in all probability, be of no value or injurious to the racial stock. I'm sure that occasionally mistakes occur as a result. But the possibility of excess and error are still no proof of the incorrectness of these laws."[7]

The Nazis particularly related German policies to U.S. Supreme Court sterilization decisions made in 1916 and in 1927 by Justice Oliver Wendell Holmes. The statements of Holmes on eugenic sterilization were favorably referred to in the German eugenics publication *Volk und Rasse* (People and race). Justice Holmes had been influenced by Davenport, Laughlin, and the ERO in his decision regarding Carrie Buck (Laughlin had presented expert testimony in the case; see chapter 7). American eugenicists had been prominent in arguing that breeding schemes could be worked out to allow the elimination of inferior racial traits.

On January 30, 1933, Adolf Hitler became chancellor of the German reich. One of the Nazi regime's first acts was to replace the Weimar Republic's voluntary sterilization law with a new mandatory act. The Nazi act, called Adverting Descendants Afflicted with Hereditary Disease, was based on the American model sterilization law. The Nazi eugenicists were particularly impressed with two American degeneration studies: those of the Jukeses and the Kallikaks.

The first of these degeneration studies had been carried out in the 1870s by prison reformer William Dugsdale. Dugsdale reputedly found that a frontiersman called Max Jukes had married a degenerate wife and produced a line of "white trash." In the Jukes pedigree, Dugsdale claimed to have found 181 prostitutes, 106 illegitimate births, 142 beggars (of these,

64 were housed at public expense), and 70 criminals, including 7 murderers. According to Dugsdale, the Jukeses had cost the State of New York over $1,308,000 for their upkeep. Ironically, Dugsdale had chosen the Jukeses as an example of how environmental conditions impact heredity. He had suggested that the burden caused by this family might have been relieved if their offspring had been placed in wholesome environments. However, Francis Galton later utilized the Jukes study as an example of hereditary degeneracy in his *Inquiries in Human Faculty* (1883). From that point on, eugenicists had begun to search for pedigrees demonstrating the inheritance of degenerate behavior. Numerous other families had been studied, none more famous than the Kallikaks, who were studied by Henry H. Goddard in 1912.

The Nazis made particular use of H. H. Goddard's study of the Kallikaks. Even more so than the Jukes study, Goddard's study had been designed to show the hereditary character of feeblemindedness. According to Goddard, Martin Kallikak sired children by two women, one a feebleminded tavern girl, the other a worthy Quakeress. From the feebleminded tavern girl came a line of criminals, morons, and other degenerates, whereas the Quakeress bore admirable human beings. The message of this study was that $150, the cost of sterilization, would have saved society the expense of all of the degenerate Kallikak offspring. Despite the fact that the studies of the Jukeses and Kallikaks turned out to be fraudulent, they survived in American psychology curricula well into the 1960s.

Indeed, American eugenicists were favorably impressed with the Nazi implementation of eugenic principles. For example, Lothrop Stoddard visited Nazi Germany in 1940, had an audience with Hitler, and served as a judge in a Nazi eugenics court. During this visit he met with many Nazi race scientists, as well as with official spokesmen of the Nazi regime. Collaboration and mutual admiration between American and European eugenicists had prepared the way for Stoddard's visit. Earlier, in 1929, Charles Davenport had presented a paper on "race-crossing" at the Second Italian Congress of Genetics under the honorary presidency of Il Duce, Benito Mussolini. We know that Davenport admired Fischer because Davenport nominated him for the presidency of the international eugenics organization. The German fascists were so eager to impress their American colleagues that Stoddard was invited to join the judges on the bench of the Eugenics High Court of Appeals. Dr. Stoddard was asked to discuss race problems and eugenics not only in Nazi Germany but also in Fascist Italy. Stoddard commented on his support for Nazi policy: "The purity of the racial strains must be preserved. . . . this is the Nazi doctrine best described as racialism." Once the Jews and other inferior stocks were annihilated, the Nazi state would be able to concern itself with "improvements within the racial stock, that are recognized everywhere as constituting the modern science of eugenics, or race-betterment."[8]

The Nazi propaganda machine made use of the support of its policies by American eugenicists. The Nazis would also specifically criticize the American treatment of Negroes, precisely when the American government questioned Germany's racial policies toward Jews. Nazi scientists and officials at the Nuremberg trials would defend their eugenic sterilization activities by citing similar laws enforced in the United States from 1907 onward. They described their work as well within established "scientific" practice. Indeed the claim has been made that the Americans even attempted to recruit some of these "scientists" for medical research after the war.[9] Although the main leader of Nazi racial philosophy and science, Alfred Rosenberg, was tried and executed at Nuremberg, other prominent Nazi eugenicists such as Fischer, Freiherr von Vershuer, and Lenz escaped prosecution. Von Vershuer, the director of the Kaiser Wilhelm Institut after Eugen Fischer, received human "materials" (such as eyes and other internal organs for his twin studies) from Josef Mengele at Auschwitz. Von Vershuer used the human organs provided from twins to estimate genetic versus environmental contributions to racial traits. After the war he was immune to de-Nazification efforts and wrote to American geneticist and Nobel laureate Herman J. Muller asking that Muller support him in his so-called troubles. Von Vershuer would later become professor of human genetics at Münster in 1951, be elected president of the German Anthropological Society, and serve as a member of the editorial board of *The Mankind Quarterly* (along with Corrado Gini, Mussolini's former "race" advisor). *The Mankind Quarterly* was founded as a postwar racial journal devoted to providing the scientific proof of Nordic supremacy. Other German members of the race hygiene movement—such as Lenz, Professor Gunther Just (once director of the Reich Health Office), and Heinrich Schade—would be similarly rehabilitated.

The historical record is quite clear concerning the intellectual support that American eugenicists gave to fascist race theory in Europe. But there was also opposition to the abuse of genetics by the fascists. Leslie C. Dunn, a leading geneticist from Columbia University, also had the opportunity to visit Germany and Italy to see eugenics in action. Dunn's report on the activities of eugenicists in Europe was influential in convincing the Carnegie Institution to distance itself from the ERO at Cold Spring Harbor. This was made increasingly easy by Harry Laughlin's continued support for the Nazis. It would not be until August of 1939 that the Seventh International Congress of Genetics in Edinburgh would officially denounce "eugenics, racism, and Nazi doctrines."[10] Unfortunately, the late date of this condemnation bears the stamp of political rather than scientific disagreement. The congress was originally scheduled to be held in Moscow in 1936. However, thirty American geneticists wrote a letter to the general secretary of the congress, Russian geneticist Simon B. Levit, requesting a special section to discuss differences among human races. In particular the group wanted to debate whether theories of racial hierarchy had any scientific basis. The Nazi government threatened to boycott the congress if this session were held. However, the Stalinists inadvertently bailed out the Nazis by canceling the 1936 congress owing to their new Lysenkoist policy against Mendelian genetics. The Stalinist biologist Trofim Lysenko held that environmental factors were more important in controlling heredity processes than were genes. Part of Lysenko's rationale for rejecting Mendelian genetics was its abuse by Western racists and classists in the eugenics movement. Soviet biology was to be reconstructed with the class interests of the proletariat in mind, and thus fascist genetics was to be rejected. Western philosophers of science have held up Lysenkoism as the case example of how not to incorporate ideological concerns into research paradigms. It is ironic that these philosophers did not notice that eugenic research practices suffered from this same flaw.

The congress finally convened in Edinburgh in August of 1939. During the conference, socialist geneticists succeeded in passing a manifesto condemning Nazi race policy. American scientists were the chief authors of this so-called *Genetico* manifesto. The activities of the ERO and its continued support for fascist race theory had polarized the American scientific community. In addition, by the late 1930s an increasing number of socialist and communist intellectuals and their sympathizers had begun to appear in the American university community. The conference manifesto called for effective birth control and the emancipation of women, stressed the importance of economic and social change, and condemned racism against ethnic minorities. The congress ended early, and German troops invaded Poland on September 1, 1939. As during the American Civil War period, academic arguments on race gave way to a life-and-death struggle in the real world.

The Logic of Eugenics and Fascism

Hitler had made it clear from the outset that the Aryan race could solve its problems only by acquiring Lebensraum, "living space," mainly at the expense of Russia. This proposition was also in concert with the racial philosophy of the Nazis and their belief in the Slavic peril. In addition, there were powerful economic incentives for the Nazis to attack the Soviet Union, for example, its invaluable oil reserves. Hitler expected the impending collapse of Bolshevism in Russia; to Hitler Bolshevism was synonymous with Judaism. The blueprint for a Nazi-ruled Europe was entirely consistent with the most flagrant forms of eugenics. Europe's resources would be used for the profit of Germany, the nation of the Aryans. The people of Europe, particularly the Slavs and the Jews in eastern Europe, would be made slaves. The undesirable elements among them (again, the Slavs and Jews and, in particular, their intelligentsia) would be exterminated. The Nazi leaders were clear in their intention to make Europe "Jew-free."

Furthermore, the Nazis had learned something about colonial domination by observing the British and French empires in Africa. The Nazis intended to destroy both the industrial capacity and the culture of the eastern European peoples. They would allow them just enough food to subsist and deny them formal education, as the United States had done to the Negroes for centuries. In fact, eugenicist Eugen Fischer would specifically comment on the plan to conscript the eastern European peoples for labor. At the conference at the Ministry of the Occupied Eastern Territories held on February 4, 1942, Fischer commented that conscript labor could not succeed if such a labor force was paid and if education was allowed in those territories. The resultant increase in the standard of living would lead to increases in the birthrate of the people the Nazis intended to eliminate.[11]

Religion would also be turned to the Nazis' favor in eastern Europe. Martin Bormann, Nazi general secretary would say: "They will preach what we want them to preach. If any priest acts differently we shall make short work of him. The task of the priest is to keep the Poles quiet, stupid, and dull-witted."[12]

Science was already fully enlisted in the ideological arsenal of the Nazis. In a lecture on June 20, 1939, referring to the "science" of eugenics and the Jewish threat, Eugen Fischer said:

> When a people wants somehow or other, to preserve its own nature, it must reject alien racial elements, and when these have already insinuated themselves, it must suppress them and eliminate them. The Jew is such an alien and, therefore, must be warded off. This is self-defense. In saying this, I do not characterize every Jew as inferior, as Negroes are, and I do not underestimate the greatest enemy with whom we have to fight. But I reject Jewry with every means in my power, and without reserve, in order to preserve the hereditary endowment of my people.[13]

Here we observe the logic of eugenics laid bare. Inferior genetic classes are made to labor for the material well-being of the superior Aryans. Exposure of inferior genetic types to forced labor and starvation drastically reduces their fertility. In fact, survival in the Nazi death camps was directly related to the ability of individuals to provide productive labor. Physicians examined aged individuals for their ability to work. Both the too-old and the too-young were exterminated. In theory, the transfer of material wealth from the "genetically inferior" should have led to the enhanced reproduction of the favored race. There is no need to refer to a detailed analysis of fertility tables to suppose that between 1933 and 1941, the height of the Nazi regime's power, successful Aryan reproduction outpaced that of the conquered peoples.

However, human society is too complex to allow for the long-term implementation of such policies. To achieve its international political and economic goals (which were, to the eugenicist, ultimately reproductive goals), the Nazi regime started a war. The result was the eventual destruction of Nazi Germany and the massive loss of Aryan lives (military and civilian). Germany lost about 6,850,000 people, equally distributed between military and civilian, or about 9.5 percent of its total population. In contrast, the Nazis inflicted about 29,737,000 deaths in the countries they targeted for racial annihilation, or about 7.15 percent of their populations.[14] Thus, while the absolute causalities were greater for the Russian and Slavic nations, the percentage of total population lost was lower. The Nazi attempt to turn the biological advantage over to the Aryans against all "biologically defective" populations in reality failed.

However, we must remember that the Nazis singled out the Jews of Europe for special attention in their extermination program. Estimates show that about 5,721,000 people, or about 68 percent of European Jews, were exterminated between 1939 and 1945. The countries of eastern Europe suffered the highest percentages of losses, with Latvia, Poland, Lithuania, Yugoslavia, Czechoslovakia, and Germany losing 89, 88, 87, 87, 83, and 83 percent of their Jewish populations, respectively.[15] Here is the real program of Galton, Gobineau, Chamberlain, Fischer, Grant, Davenport, Laughlin, and company. The Nazis put into practice what these "scholars" of race hygiene merely preached.

CHAPTER 9

The Retreat and Postwar Revival of Scientific Racism

World War II may have had more impact than any preceding historical event in creating an atmosphere that allowed scientists to question theories of racial hierarchy. The war's impact resulted in part from the fact that the Axis racial philosophy was a core component of its rationalization for war. Hitler had promised that he would exterminate European Jewry. The Italian Fascists glorified themselves in the lost racial honor of the Roman Empire and began their attempt to vindicate their race by invading Ethiopia. The Japanese championed themselves as the "master race" of Asia and demanded their place as Asia's "racially" appropriate rulers, as opposed to the Americans, English, Dutch, and French. Soldiers from Germany, Italy, and Japan went to war amid political rhetoric and pseudoscientific propaganda that pronounced their racial destinies to rule the world.

Conversely, African Americans, in the main, rejected fascism and racism from the beginning. Many African Americans served and died in the international brigades in Spain. However, the war against fascism heightened African Americans' consciousness of their own position in the United States as an oppressed racial minority. As early as 1937, the NAACP approached President Roosevelt about the concerns of African Americans in the coming war. The NAACP warned that it would be difficult for African Americans to support the war unless the United States attended to their domestic grievances. The threat of a labor march on Washington forced Roosevelt to sign Executive Order 8802, ending discrimination against African Americans in the defense industry. African American leaders continued to plead to Roosevelt to integrate all branches of the military, including the officer corps. Nevertheless, the armed forces retained segregation in all army units. In fact, the command of the American forces actually stood by prejudice against African American soldiers in combat. In 1943 the secretary of war banned a pamphlet entitled *The Races of Mankind*, by Gene Weltfish and Ruth Benedict, because it challenged the myth of white intellectual superiority.[1] African Americans suffered other ridiculous insults throughout the war, such as the army's insistence on maintaining segregated blood supplies, this despite the fact that Karl Landsteiner had already demonstrated that the basis for blood incompatibility—the proteins produced by the A, B, O locus—had nothing to do with race. Furthermore, the very technique to store blood for future transfusions had been pioneered by the African American physician Charles R. Drew.

Despite all the hardships African Americans faced before and throughout the war, they fought heroically on all fronts. For example, African Americans provided the first symbolic victories over fascism: Jesse Owens's victories in the 1936 Berlin Olympics and Joe Louis's defeat of German boxer Max Schmeling in 1938. These victories were crucial because Nazi race theory preached the complete supremacy of the Aryan/Nordic race in all fields of human endeavor, intellectual, moral, cultural, and physical. Although originally opposed to a Schmeling-Louis fight, the Nazis viewed Schmeling's victory in an earlier contest (1936) as a validation of German racial superiority. In 1938 Joe Louis would redeem himself with a one-minute, first-round knockout of Schmeling. A month after the first Schmeling bout, the myth of Aryan/Nordic physical superiority was shattered in Berlin when Jesse Owens, a gifted African American track star from Ohio State University, won four gold medals in the 1936 Olympic Games. Owens won the 100-meter dash and the running broad jump, equaling the Olympic record; and he set new Olympic and world records in the 200-meter dash. Finally, he was added to the 400-meter-relay team, which also set new Olympic and world records. Sadly, he and John Woodruff had replaced the Jewish members of the team, Marty Glickman and Sam Stahler, in part not to further infuriate the Nazis. Adolf Hitler refused to acknowledge Owens's victories. It is said that after Owens's victory in the 100 meters on the second day, Hitler stopped personally congratulating the athletes.[2] Once again, however, the United States demonstrated its hypocritical racial attitudes. Upon returning home, Jesse Owens never received the promised commercial endorsements resulting from his Olympic victories. At one point he was reduced to participating in exhibitions in which he raced against horses for public amusement.

African American soldiers served gallantly in the war; the combat records of the 761st Tank Battalion and the Tuskegee Airmen (99th Pursuit Squadron) could not be ignored. Still, in World War II most African Americans in the military were relegated to service and support roles. Dorie Miller, an African American who won the Navy Cross for his heroism aboard the USS *West Virginia* at Pearl Harbor, was still a messman when he died aboard the USS *Liscome Bay* in 1943. Despite ongoing discrimination, the full participation of African Americans in the war effort, both abroad and at home, raised their expectations for full participation in postwar American society. Indeed, by the end of the war Euro-Americans were beginning to realize that they could not expect to keep African Americans entirely disenfranchised. In an attempt to win the 2.5 million African American votes in the 1944 election, the Republicans issued a new platform promising to establish a fair employment office, to provide fair housing, to enact legislation against lynching, to investigate the mistreatment of Negroes in the armed forces and enact corrective legislation, and lastly to abolish the poll tax by constitutional amendment. The end of the war and the United States' new international relationships were forcing American political leaders to realize that leadership of the "free" world was ideologically inconsistent with segregation and Jim Crow at home. The country could not stand against the spread of communism while maintaining flagrant racism in its own society. Furthermore, the Nuremberg trials had revealed for the world the relationship between racism and fascist ideology. The revelations of the German death camps made it impossible for anyone to defend racist public policies based on eugenics, at least not without ridicule and denunciation.

The Retreat of Scientific Racism

The retreat of scientific racism had its beginnings in the 1930s. We have seen how antifascist sentiment developed at that time among American scientists, who were particularly critical of the antidemocratic rhetoric that the classical eugenicists had supported. On the eve of World War II, Franz Boas was able to organize the publication of a statement by American scientists rejecting Nazi race theory, with 1,284 signatories, including 3 Nobel laureates and 64 members of the National Academy of Sciences. The statement was published in the *New York Times* on December 11, 1938. Earlier that same year, Boas's disciple Ruth Benedict had published *Race, Science, and Politics,* a volume devoted to dismantling racist concepts in science and outlining their relation to fascism. Later, during the war, the American Association for Physical Anthropology would condemn the segregated blood banks of the American army on the basis that the practice mirrored Nazi race theory.

Blood studies done after the war would further erode the classical conceptions of race. For example, electrophoretic studies of hemoglobin, the protein involved in sickle cell disease, beginning in the 1940s and continuing into the 1950s would reveal that there were many hemoglobin variants in the world's populations. (By 1931 Linus Pauling had already begun to suspect that sickle cell disease was a protein polymorphism disease.) Serological data obtained by A. S. Wiener in 1946 would contradict Ashley Montagu's grouping of the Australoids with the Negroids in his 1945 taxonomy based on anthropometric data. Negroids and Australoids might share the same pigmentation or skull-shape pattern and yet be entirely different in blood allele frequencies. Thus, the more that data became available from the direct examination of genetic variation in humans, the more that the classical race concept unraveled. There was too much variation within populations, too little difference between them.

Population Genetics and Racial Ideology

In the 1930s important developments in population genetics began to weaken the classical views of biological variation and its application to eugenics. By 1930, Sir Ronald A. Fisher had worked out what would be called the fundamental theorem of natural selection, which stated that the rate of increase in fitness for an organism was equal to its genetic variance in fitness. This principle showed that genetic variation was the raw material of evolution. By contrast, the classical theory of natural selection saw all mutations, the ultimate source of genetic variability, as deleterious; and eugenicists had used the theory to argue against the value of genetic variability.

Through the 1940s, data began to accumulate favoring the "balance" theory of population genetics, which holds that situations in which heterozygotes with an advantage over either homozygous genotype might account for the maintenance of genetic variation. Compared to the classical selectionist model, which predicts that one favored allele will dominate and a second deleterious allele will be eliminated (see figure 7.3), the balance theory predicts that both alleles will be maintained at intermediate frequencies determined by the relative fitness of the three genotypes (see table 9.1 and figure 9.1).

The relationship between sickle cell disease and malaria provides an example of heterozygote advantage. The frequencies of the allele for sickle cell disease are high in populations exposed to malaria because being heterozygous for the sickle cell protein confers resistance to the malaria parasite. In environments where malaria is endemic, individuals who are heterozygous for the sickle cell protein receive protection. Individuals who are homozygous for the normal protein face reduced fitness due to the malaria parasites, and individuals homozygous for the sickle protein face reduced fitness due to sickle cell disease. Contrary to popular belief, sickle cell disease did not originate in western Africa, nor is it a black disease. It

Table 9.1. General Model for Heterozygote Advantage

	Dominant	Heterzygote	Recessive
Genotype	AA	Aa	aa
Fitness	$(1-s_1)^a$	(1)	$(1-s_2)^a$

[a]s_1 and s_2 are the selection coefficients against the homozygous dominant and recessive genotypes, respectively.

is found at high frequency in many populations in conjunction with the presence of malaria (see also chapter 11).

The connection between sickle cell disease and malaria is also an example of the environmental determination of genotype fitness. In environments where malaria is uncommon, such as the United States, this example becomes the same as the classical selection against a recessive gene (fitness of AA = 1, Aa = $1 - s_1$, and aa = $1 - s_2$, where s_1 = selection against individuals heterozygous for the sickle cell trait and s_2 = selection against individuals homozygous for the sickle cell trait). That is because the fitness of homozygous normal individuals increases when they are not exposed to mortality from malaria. Also under these conditions we expect the mortality of heterozygous individuals to be less severe than those carrying two copies of the sickle cell allele. This is because heterozygous blood cells are only partially sickled, whereas the blood cells of homozygous individuals are severely deformed. If heterozygote advantage and environmental determination of fitness were common, the eugenicists would also be wrong in championing the reduction of allelic diversity.

Heredity, Race, and Society, 1946

In 1937, the Russian-born American geneticist Theodosius Dobzhansky, who worked in Thomas H. Morgan's research group at Cal Tech, realized that there was much more genetic variation in nature than the classical theory could accommodate. In 1939 Dobzhansky left Cal Tech to accept an appointment at Columbia University. There he became a colleague of the anthropologist Franz Boas and the geneticist Leslie Dunn. The three men participated in an organization called the Columbia University Federation for Democracy and Intellectual Freedom. The goal of this group was, in part, to show that genetic science did not support totalitarian social regimes.

In their 1946 work titled *Heredity, Race, and Society*, Dunn and Dobzhansky pointed out that the historical schemes of race classification failed to uniquely identify individuals as to race. Antiquated systems such as Lin-

naeus's and Blumenbach's schemes of emotional and morphological classi-
fications were clearly inadequate. Dunn and Dobzhansky also identified
(as had Darwin) the racial multiplication problem: that is, as the number
of phenotypic traits used to classify races increased, so too did the number
of races. In 1889 J. Deniker identified twenty-nine races; and in 1933
E. von Eickstedt identified three major races, eighteen sub-races, three col-
lateral races, and three intermediate "forms." By 1950, Carleton Coon,
Stanley Garn, and J. Birdsell would identify six putative stocks and thirty
races.[3] The disparity of such efforts resulted from scientists' inability to
agree on what criteria should be used to constitute a race. The inclusion of
cultural features, such as language, made the problem worse because now
national borders did not uniquely identify racial groups either. There were
no American, French, English, or German races. A simple explanation of
this failure to match culture and race was that historical biological variation (both genetic and environmental) did not follow socially constructed
modern nation-state borders.

In *Heredity, Race, and Society,* Dunn and Dobzhansky also proceeded
to identify the chief source of confusion between scientific and lay attempts to classify races. For example, by using segregated populations in
major American cities such as New York (Harlem, Little Italy, the Norwe-
gian colony in Brooklyn), they explained that a lay observer could easily
distinguish physical features (differences in hair type and color, nose
shape, eye color, and so on) and then assign these populations to races,
such as those used by anthropologists (Negro, Nordic, Alpine, and
Mediterranean). One general problem encountered once these groups are
identified is the tendency to assume that physical features correlate with
mental and cultural ones. Furthermore, although these populations differ
in mean physical characters, no single individual may be uniquely identified by the group average in characters. Negroes and Nordics might be
relatively easy to separate. However, many errors would ensue from attempting to assign Nordics, Alpines, and Mediterraneans. The problem
becomes greater when populations that originate in closer geographical
areas are compared.

Most importantly, Dunn and Dobzhansky critically examined the concept of "pure" race. To adequately understand why and how races can be
considered pure, we must first understand what races were thought to be
and how the idea of them came into existence. In neo-Darwinian theory,
the term "race" is best applied to a geographically isolated or localized
subpopulation. Isolation is important because it prevents immigration or
emigration, thus eliminating the flow of genes between localities. Under
these conditions, gene frequencies in the isolated population may through

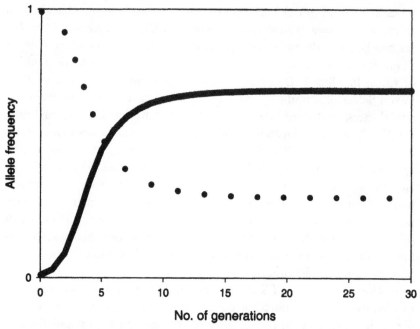

Figure 9.1. Equilibrium frequency for heterozygote advantage: • • • • = allele A, ——— = allele a. This figure illustrates heterozygote advantage during natural selection. If the heterozygote (genotype Aa) has greater Darwinian fitness than either homozygous genotype (AA, aa), then the allele frequency (p for gene A and q for gene a) will equilibrate at intermediate frequencies. The higher fitness of sickle cell allele heterozygotes in malaria zones is an example of this principle.

time become very different either because of natural selection under local conditions or because of genetic drift.

There are a number of mechanisms by which a population can become isolated, for example, geographical barriers such as mountains or rivers. Charles Darwin and Alfred Russel Wallace studied naturally isolated populations (for example, those on the Galapagos Islands and the islands in the Amazon River) to develop their theories of natural selection. Simple geographic dispersion can also isolate populations. In this situation, one might expect gene frequencies at the one end of geographic traverse to be quite different from those at the other, even though the change from any one point to the next along the traverse would be imperceptible. Consider, for example, the gene frequencies one might encounter between Egypt and Mongolia. A collection of gene frequencies at any particular locus in these two locations would reveal rather different alleles predominant. However, there would be continuous gradation between the two localities.

The scenario of geographic dispersion is particularly illustrative in terms of the historical development of human genetic diversity. In both models of the evolution of modern humans (the recent origin, African Eve hypothesis and the polygenic origin hypothesis), geographical distance and physical barriers preventing gene flow between populations play a major role in producing genetic differentiation. Thus when one examines gene frequencies, one must ask, Where does one race end and another begin? The existence of pure races is intimately tied to an unambiguous statistical means to delineate such races. We now know that collections of human genetic data reveal no such means to clearly delineate races.[4]

The difficulty of identifying pure races on the basis of gene frequencies is exemplified by A, B, O blood-group polymorphism in humans (see table 9.2).[5] We could use the data in table 9.2 to reassign various populations into "O" races by starting, for example, at 25 percent O frequency and creating 5-percentage-point groups (see table 9.3). It should be clear from this example that the assignment of these groups is completely arbitrary: we could reassign them by starting at a different percentage and using a different increment.

In addition it is possible to use other loci and draw entirely different racial groups. For example, we could regroup populations according to the degree of resistance to malaria conferred by sickle cell hemoglobin. There are a number of other single-gene traits we could use: taste sensitivity to phenylthiocarbamide, color blindness, deficiency in the enzyme g6pd, MN blood groups, β-amino isobutyric acid secretion, Tay-Sachs allele, or Paget's disease. None of these loci yields the racial groupings assigned by classical anthropology. Clearly, single-gene markers do not allow a reasonable classification of human diversity. Intuition suggests that the use of genetic markers to define races would become more accurate if more loci were used, but even then there are still no unambiguous ways to recover the classical anthropological categories (see chapters 10 and 11).

This reasoning allowed Dobzhansky and Dunn to illustrate the major fallacies of racial classification in humans. Dobzhansky would also point out that it was a mistake to treat as single units the complexes of characteristics of individuals, races, and species. Any attempt to find rules governing the inheritance of such complexes would be fallacious. What was important was the inheritance of separate traits, not complexes of traits. Therefore a key error in the work of modern racial thinkers was their failure to see that genetic traits were distributed independently among various populations. The independent inheritance of genetic traits in populations results from the fact that physical and biological gradients responsible for creating selection pressures vary independently. This is

Table 9.2. Frequency of the O Allele at the A,B,O Locus, by Population

Population	O allele frequency
Europeans	
Icelanders	56
Scots	54
English (from the south)	43
Spaniards	42
Norwegians	40
Swedes	38
Finns	34
Sicilians	46
Russians	32
Africans	
Egyptians	27
Ethiopians	38
Pygmies	31
Nigerians	57
Asians	
Tartars	28
Kirghiz	32
Buriats	32
Chinese	31
Japanese	30
North Americans	
Eskimos	41
Navahos	75
Blackfeet	24
South Americans	
Toba	97
Australians	
Aborigines	48

SOURCE: These data are cited from L.C. Dunn and Th. Dobzhansky, *Heredity, Race, and Society* (New York: Mentor Books, New American Library, 1946), 120.

known as the principle of discordance. There is no reason for alleles found at a locus that determines skin color, hair type, or disease resistance to be found consistently associated among populations. As I have said, Sri Lankans of the Indian subcontinent, Nigerians, and Australoids share dark skin tone but differ in hair type and genetic predisposition to disease.

Table 9.3. Reassignment of the Populations, by O Allele Frequency

O allele frequency	Populations
25–30	Blackfeet, Egyptians, Japanese, Tartars
31–36	Buriats, Chinese, Finns, Kirghiz, Pygmies, Russians
37–42	Eskimos, Ethiopians, Norwegians, Spaniards, Swedes
43–48	Aborigines, English, Sicilians
49–54	Scots
55–61	Nigerians, Icelanders
> 61	Navahos, Toba

The UNESCO Statements on Race

On July 18, 1950, the *New York Times* ran a front-page article, under the headline "No Scientific Basis for Race Bias Found by World Panel of Experts," reporting on the United Nations Educational, Scientific, and Cultural Organization (UNESCO) statement on race. The preamble to the statement of the UNESCO committee read

> Racial doctrine is the outcome of a fundamentally antirational system of thought and is in glaring conflict with the whole humanist tradition of our civilization. It sets at nought everything that Unesco stands for and endeavours to defend. By virtue of its very Constitution, Unesco must face the racial problem: the preamble to that document declares that "the great and terrible war which has now ended was a war made possible by the denial of the democratic principles of dignity, equality and mutual respect of men, and by the propagation, in their place, through ignorance and prejudice, of the doctrine of the inequality of men and races."[6]

The following is an excerpt from the body of the statement:

> From the biological standpoint, the species *Homo sapiens* is made up of a number of populations, each one of which differs from the others in the frequency of one or more genes. Such genes, responsible for the hereditary differences between men, are always few when compared to the whole genetic constitution of man and to the vast number of genes common to all human beings regardless of the population to which they belong. This means that the likenesses among men are far greater than their differences.[7]

The 1950 statement can be seen as a real advance in promulgating antiracist ideas. For example, the prewar efforts to draft such a statement by the Institute of International Cooperation (which contained many of the same scientists) had to be abandoned so as not to antagonize Hitler. Nazi scientists who had escaped responsibility for war crimes were quick to denounce the 1950 UNESCO statement. Hans Gunther, Eugen Fischer, and Fritz Lenz continued to discuss racial theory from the point of view of eugenics and Nordic superiority. Lenz said that the statement "runs counter to the science of eugenics." He further added that "psychical hereditary differences between the races are more important than the physical differences." He still maintained that Jews were a race distinguished by the former.[8]

UNESCO felt the need to issue a revised statement on race a year after the first, in 1951. The 1951 statement indicates that there was not yet uniform agreement about the invalidity of the race concept. The new statement, drafted by a group of physical anthropologists and geneticists, claimed the following:

> The concept of race is unanimously regarded by anthropologists as a classificatory device providing a zoological frame within which the various groups of mankind may be arranged and by means of which studies of evolutionary processes can be facilitated. In its anthropological sense, the word "race" should be reserved for groups of mankind possessing well-developed and primarily heritable physical differences from other groups. Many populations can be so classified but, because of the complexity of human history, there are also many populations which cannot be easily fitted into a racial classification.[9]

This statement seems to represent both a scientific and a political retreat from that of 1950. It seems to admit that racial classification schemes do not work well for humans, yet it still defends the validity of the concept of race. This problem might have arisen from the use of the term "heritable" in the definition of the word "race." Heritable differences ultimately refer to genetic features of populations. However, at this point little was known about genetic variation in human populations. The best data were those related to protein polymorphism, but even these data had already begun to shed serious doubt on the validity of the race concept.

The members of the 1950 and 1951 committees differed notably in their national origin, ethnicity, and scientific discipline.[10] The 1950 committee had a total of twenty-one drafters and reviewers. None came from the former Axis countries; three members came from former colonies; one was African American (historian E. Franklin Frazier); and several were humanists (Frazier, sociologist Gunnar Myrdal, and psychologist Otto Klineberg). In contrast, the 1951 committee was composed of fourteen anthropologists and geneticists from the United States and Europe. Only Dahlberg, Dunn, Montagu, Dobzhansky, and Huxley participated on both committees. It is likely that the 1951 document resulted from dissatisfaction among more scientifically conservative members of the genetics and anthropology communities, who might have believed that liberal "political" concerns had compromised the scientific integrity of the first statement. Conversely one can argue that the absence of the humanists and minorities from the second committee was itself a political statement.

Despite its weaknesses, the 1951 UNESCO statement clearly assaulted one of the key tenets of Nazi racial theory, the existence of consistent mental differences among human racial groups. The statement pointed out that most anthropologists did not include mental characteristics in their classification of human races. It is precisely the question of whether race and intelligence were linked that would be brought into sharp contention in the battle over school integration in the United States.

"With All Deliberate Speed"?

It is likely that the UNESCO statements on race had little influence on the racial climate of the United States in the 1950s. African Americans were prepared, with or without scientific backing, to make an assault on school segregation. The South as a whole spent three times more money per white student than per black one, and states like Georgia and Mississippi, five times as much.[11] In 1944 the seventeen states that segregated school children spent over $42 million on busing white children to school but only $1 million on black children. Of course, without transportation black children would be less likely to finish their education. Although the NAACP had won a major victory in the 1938 *Gaines v. Canada* decision, which had outlawed discrimination in higher education, the South still spent $86 million and $5 million on white and black public universities, respectively. There were major disparities in the types of professional training one could receive. There were no historically black colleges or universities (HBCUs) where one could receive a doctorate; there was one medical school for blacks and twenty-nine for whites. The same inequities were found in the numbers of white and black professional schools of pharmacy, twenty to one, law, forty to one, and engineering, thirty-six to zero. In addition, one-fourth of African Americans in the South during this period were functionally illiterate. Clearly, segregated education was performing its function of keeping African Americans virtually powerless against Euro-American oppression.[12]

After a series of successful challenges to segregation in higher education, the NAACP decided to take up the *Brown v. the Board of Education of Topeka* case. The case had originally been brought before the Supreme Court, under Chief Justice Frederick Vinson, in 1952. President Harry Truman had shown his support for the equality agenda of *Brown* by allowing the Justice Department to file an *amicus curiae* brief. In the brief, the federal government emphasized the foreign policy implications of racial segregation, suggesting that segregation supported communist propaganda and that foreign visitors mistaken for American Negroes were being refused food, lodging, or entertainment. The philosophically divided Vinson court postponed argument until June 1953. Vinson died before the reargument could commence, and the new president, Dwight Eisenhower, appointed Earl Warren as the new chief justice. The Eisenhower administration would not give clear support to the *Brown* plaintiffs, fearing loss of support for the Republican Party in the South, but would not openly oppose civil rights. In 1954, the Warren Court decided for the plaintiffs on the basis of its determination that segregation caused significant psychological harm to African American children and thus violated the equal protection clauses of the Fourteenth Amendment. The determination that segregation harmed African American children rested in part on interpretations of psychological and sociological data presented by the NAACP's expert, Kenneth Clark (for example, the doll test, cited in footnote 11 of the Court's decision). To test the self-images of African American children, Clark had asked children a series of questions regarding white and black dolls. From the responses to such questions as "Which is the nice doll?" Clark had concluded that segregation was harming African American children.

In the years following the *Brown* decision, the NAACP's use of this ostensibly scientific argument would be crucial in suggesting that equal protection under the law was somehow conditioned on the interpretation of scientific evidence bearing on the impact of public policies. Indeed, southern racism soon began to raise the specter of "interracial mongrelization" as the only possible result of school integration. Conservative southern citizens groups began to organize academic and extralegal resistance to integration. The governor of Alabama commissioned a study of the "race" problem, which warned southerners about "protoplasmic" mixing of the white and Negro races. This fear of integration throughout the United States was a major catalyst in the resurgence of scientific racist ideology.

The Revival of Scientific Racism in Defense of Segregation

One of the first southern "scientists" to denounce the 1954 *Brown* decision was Carleton Putnam. Putnam insisted that desegregation had been the result of an antiscientific movement that denounced the truth of hered-ity in favor of egalitarian environmentalism. Putnam felt that southern re-sistance to desegregation should be based on "innate black inferiority." Another racist voice denouncing integration was that of Ernest van den Haag. He was critical of Kenneth Clark's improper use of the scientific method in the doll test, and he supported segregation and the denial of voting rights to blacks in South Africa and the United States. He argued that the experience of being shunned by white classmates in integrated schools would be far worse psychologically for blacks than segregation.

In another attempt to use science in the service of segregationist arguments, intelligence testing was resurrected shortly after the *Brown* de-cision. The segregationist argument against integration was that inferior black performance in schools and in measurements of general intelligence was genetically predestined rather than the result of a segregated learning environment or of social discrimination. In 1956 Frank McGurk, a pro-fessor of psychology at Villanova University, reviewed six studies of test performance carried out between 1935 and 1951. His review appeared in *U.S. News and World Report,* a magazine that throughout this period consistently ran prosegregationist articles. McGurk found that there had been no reduction of the black-white differential in intelligence test scores as compared to the World War I army studies, despite supposedly substan-tial improvements in social and economic opportunities for blacks.

The psychologist Henry Garrett, who had been chair of psychology at Columbia University while Kenneth Clark was a graduate student, would become the chief "scientist" opponent of integration. Garrett had used moderate and scientific language before *Brown;* however, he soon di-gressed into pure racist rhetoric. In a 1962 letter to *Science,* Garrett wrote: "No matter how low . . . an American white may be, his ancestors built the civilizations of Europe, and no matter how high . . . a Negro may be, his ancestors were (and his kinsmen still are) savages in an African jungle."[13] Garrett also embraced the White Citizens' Council movement, which urged white citizens to boycott African American businesses.

Garrett wrote worse things in *The Mankind Quarterly,* which had been founded in Edinburgh in 1960 by individuals who believed in the innate superiority of the white race. Most of the founders were members of avowedly racist political organizations. The founders included Robert Gayre, a Scottish physical anthropologist with deep ties to neo-Nazi orga-nizations; Garrett; and even Corrado Gini, Mussolini's race advisor and author of *The Scientific Basis of Fascism.* Gayre considered himself a Strasserist Nazi (the Fascist faction that had opposed Adolf Hitler). Others in *The Mankind Quarterly* orbit included Roger Pearson (who would later become president of the pro-KKK Liberty Lobby and translator of Hans Gunther's works on Aryan religion) and the renowned British racist R. Ruggles Gates (the British Davenport).

The organization behind *The Mankind Quarterly* was the International Association for the Advancement of Ethnology and Eugenics (IAAEE). The IAAEE's membership included the same individuals who made up the journal's editorial board, for example, Garrett and physicist and eugenicist William Shockley. Henry Garrett was also a director of the Pioneer Fund. Founded by the Massachusetts textile magnate Wickliffe Draper, the Pioneer Fund included among its original directors eugenicist Harry H. Laughlin and financier Frederick H. Osborn, the nephew of Henry Fairfield Osborn. Throughout its history it has been a major source of funding for eugenic research, purportedly to improve the human race. It supported much of the research that was relied upon by *The Bell Curve* to make its case for the genetically-based superiority of East Asian/Caucasoid, as opposed to African, intelligence.

Less than twenty-five years after Auschwitz, the scientific racist infrastructure was firmly reestablished. Although the UNESCO statements of 1950 and 1951 had declared that race was not a useful concept in human biology, and that anthropologists generally should not utilize intelligence in racial classification schemes, race and intelligence would dominate American thought concerning integration.

We should also remember that the 1950s were a period of revolution and fierce resistance in American race relations. The *Brown* decision was not even six months old when Rosa Parks's refusal to give up her bus seat to a white man in Montgomery, Alabama, led to the Montgomery bus boycott of 1955–1956. The boycott thrust the Reverend Martin Luther King Jr. onto the national scene as a civil rights leader. His first reward for this notoriety would be the bombing of his home on January 30, 1956. By 1960 the Student Nonviolent Coordinating Committee would be formed after a wave of student sit-ins at segregated lunch counters all over the South. It was now impossible to avoid the politicization of any topic having to do with race in the United States.

Biological Theories of Race at the Millennium

In the preceding sections, we have seen the origin and transformation of biological conceptions of race. The chief agent of change in these conceptions was Darwinian evolutionary theory. This theory was born incomplete, and its initial applications to the problem of race suffered from that incompleteness. In the 1950s, the discovery of DNA and the subsequent understanding of the mechanisms that governed heredity had a profound impact on this problem. The invention of new molecular techniques for studying genetic diversity at even finer levels of resolution made possible the growth of new evolutionary theory.

In the 1960s, the classical and balance theories of population genetics were joined by the neutral theory of the Japanese geneticist Motoo Kimura. This theory proposed that many genetic substitutions at the DNA level led to the formation of protein variations having no impact on an organism's fitness. Thus, the DNA of a population would accumulate many more changes than had been predicted by previous theories. Determining the validity of the neutral theory would require the testing of genetic diversity in a variety of organisms at the molecular level, and by the 1980s a significant body of literature had been produced on the amount of genetic diversity that actually existed within and among the populations in the human species. Scientists soon realized that the largest amount of genetic variation in our species resides at the level of the individual rather than the group. In 1982 Masatoshi Nei and Arun K. Roychoudhury stated: "Coon . . . identified five subspecies of man, i.e., Caucasoid, Mongoloid, Congoid (Negroid), Capoid (Bushman and Hottentot), and Australoid. However, the genetic distances for protein loci between Caucasoid, Mongoloid, and Negroid are of the same order of magnitude as those for local populations in other organisms and considerably smaller than those for subspecies. . . . Therefore it is not appropriate to assign the rank of subspecies to the major races of man."[1]

Soon, the vast majority of molecular and physical anthropologists would arrive at the conclusion that the classical racial categories were simply not useful in describing the human species. On Tuesday, February 21, 1995, several major newspapers ran the following headline: "No Such Thing as Race,

Genetic Studies Say."[2] The recent work examining human genetic variation and the race concept was the subject of at least two symposia at the 1995 annual meeting of the American Association for the Advancement of Science. The panels for both of these symposia were composed of philosophers, anthropologists, biologists, and social scientists. They concluded that although the term "race" had no biological validity, it was, however, still an important operational social and cultural category.[3] In 1998, *Science* magazine's genome issue devoted a section to the rift between genetic reality and the racial categories used by the Office of Management and Budget (OMB) for the 2000 census. Yale geneticist Kenneth Kidd stated: "One of the benefits that's going to come from [studies of genome diversity] is an even greater understanding of how similar we all are in our marvelous variation."[4]

Despite the unambiguous character of the recent studies of human genetic diversity, the significance of these results for our understanding of socially constructed races has not been fully appreciated. The fact that no races exist in our species has not been adequately communicated to the lay public. Part 4 of this work examines two fields of study, intelligence testing and biomedical research, where the race fallacy still holds considerable influence.

CHAPTER 10

The Race and
IQ Fallacy

No issue in the history of racial theory has been more pernicious than the idea that the races within the human species differ significantly in their innate intelligence. The proposition that there are genetic differences in intellect and temperament is as old as the race concept itself, but no real objective means to test this proposition has ever existed. Eighteenth-century naturalists such as Linnaeus classified the human races by vague criteria such as customs, law, and temperament; and the nineteenth-century polygenists would substitute reputedly objective criteria such as cranial volume and skull angles. These men, of course, assumed a direct link between those morphological features and the intellectual capacities of the races. As we have seen, the nineteenth-century theorist Joseph Arthur, comte de Gobineau invoked the ability of a race to found a great civilization. However, European archaeologists could never seem to locate any great African civilizations. Even Sir Francis Galton, in his *Hereditary Genius,* did not advance an accurate or objective means to measure intellect, yet he managed to rank the races from the ancient Greeks to modern Negroes (see chapter 6).

The twentieth-century psychometric research program stands alone in its effort to create objective measures of cognitive ability and then relate those to genetically determined racial variation. In 1905 Alfred Binet, director of the psychology laboratory at the Sorbonne in Paris, devised a series of tests to investigate the rates at which different individuals "learned." He utilized a series of questions that were graded by the age at which the individual was able to correctly answer the question. The average score of the group was used to determine what was "normal." This score was considered the "mental age." In 1912, German psychologist William Stern created the intelligence quotient (IQ), by suggesting that an important comparison could be made by dividing the mental age by the chronological age and multiplying the result by one hundred.

Binet's purpose in constructing his tests was to help identify children who needed special educational assistance in the "normal" classroom. The metric was designed therefore to be a gross indicator of normal versus subnormal performance in learning. Binet never intended it to be used to label or stigmatize children, nor did he intend it to be used to create racial scales of intelligence. Unfortunately, this is precisely what happened in 1908, when Henry H. Goddard (the chronicler of the Kallikak story, chapter 8) translated the tests for use at his Vineland, New Jersey, Training School for Feeble-Minded Girls and Boys.

Thus began the American school of psychometry. This research program was launched with a series of core principles, principles that in reality rested on a series of untested assumptions. These principles can be summarized as follows:

1. Every individual had an underlying general intellectual capacity.
2. This capacity could be accurately tested by means of standardized tests.
3. Variation in individual intellectual capacity was mainly innate (determined by genes).
4. An individual's genetic intellectual capacity was related to his or her racial origin.
5. The intellectual and reproductive capacities of races were inversely correlated.
6. The differential reproduction or immigration or both of inferior races was lowering American intelligence (dysgenesis).

These assumptions would play a crucial role in some of the most infamous conclusions of early American psychometry. These included the labeling of certain European immigrants as unworthy of assimilation; the declaration that IQ test scores made the United States unsafe for democracy because the average person was too moronic to make important decisions; and the claim that IQ test score differences between southern and northern Negroes were due to differential migration of intelligent genotypes. We have already seen how the army IQ tests were used to help bias immigration quotas in the 1920s. Indeed, even the Scholastic Aptitude Test (SAT) was originally invented to control Jewish entry into higher education.

Eventually, both Goddard and Lewis Terman (inventor of the Stanford-Binet test) would recant the hereditarian theory of IQ late in the 1930s.[1] The United States would nevertheless continue to rely on the IQ concept through the 1940s, but the coming of World War II and the subsequent decline of the eugenics movement mitigated major controversies during this period. Genetically determined racial theories of intelligence were resurrected as a core component of the postwar revival of scientific racism, par-

ticularly in response to the problem of school integration. The chief pro-
tagonist in this revival was Arthur Jensen.

Paul on the Road to Damascus:
Is There Danger of Genetic Enslavement?

No one better typifies the return to scientific racist ideology in the period
after World War II than eugenicist Arthur Jensen. It is said that the apostle
Paul, while on the road to Damascus, was struck by a vision that accomplished his conversion to the true faith. So it was with Jensen's conversion.
Jensen did not start out as a racist ideologue. In a volume published in
Great Britain in 1961, Jensen wrote that the lower socioeconomic status
of Negroes and Mexicans could not be interpreted as evidence of their genetic inferiority. He pointed instead to the powerful racial barriers to up-
ward social mobility. Jensen stated that the historically low IQ scores of
the Negro population were most likely due to environmental rather than
genetic differences. Yet by 1969, Jensen had changed his mind. In that
year, the then relatively unknown psychologist would pen his infamous
Harvard Educational Review article titled "How Much Can We Boost IQ
and Scholastic Achievement?" which became the manifesto of the modern
psychometric movement. Its message was simple: "The differential birthrate, as a function of socioeconomic status, is greater in the Negro than in
the white population. . . . Much more thought and research should be
given to the educational and social implications of these trends for the future. *Is there danger that current welfare policies, unaided by eugenical
foresight, could lead to the genetic enslavement of a substantial segment of
our total population?*" [italics added].[2]

The change in Jensen's viewpoint takes on particular significance in
light of the civil rights movement. The movement's early program had focused on dismantling legal props to segregation. The Civil Rights Act of
1964 and the Voting Rights Acts of 1965 had essentially achieved that
end. However, the passage of these acts had not addressed the ongoing denial of economic opportunity to African Americans. Even leaders like Dr.
Martin Luther King Jr. recognized these acts were not enough and began
to change the focus of their efforts. These fundamental contradictions between the rhetoric of the movement and the realities of black life exploded
in the summer of 1967, which saw some of the worst racial disturbances
in the history of the United States. President Lyndon Johnson's National
Advisory Committee on Civil Disorders examined the underlying causes
of the civil unrest. The committee (also called the Kerner Commission
after its head, Illinois governor Otto Kerner) investigated genetic and

biological explanations, among others, for the violence. The committee concluded, however, that white racism was the principal cause of the disturbances.

The Euro-American intellectual community did not unanimously accept the Kerner Commission's indictment of institutional racism. Some scientists were quick to provide analysis in support of white supremacy and segregation. Robert Kuttner (of the IAAEE) pointed out in the respected scientific journal *Perspectives in Biology and Medicine* that the feasibility of changing the position of any minority group rested on the assumption that its genetic capacity was roughly equal to that of the majority. This statement reveals the core argument of the modern psychometricians and eugenicists concerning the racial structure of society: discrimination is not the reason for the failure of minority races in the United States; instead the relative social positions of individuals are fairly determined by their intellectual capacities, which are genetically determined. If society shows a consistent pattern of white domination, this is because the white race has a higher number of genetically favored individuals. Thus, it is the genetic inferiority of the Negro, not the action of the whites, that accounts for his degraded condition.

At the turn of the twentieth century, the proponents of this school of thought had seen the genetic inferiority of Negroes as general, and they had believed that the low fecundity of Negroes and their greater mortality rate would guarantee their extinction in North America. These beliefs account for the eugenicists' greater concern with the immigration of inferior races from Europe. However, changes in American society in the early twentieth century began to improve the overall health and thus the reproductive fitness of African Americans. The 1960s were the first decade in which African American populations increased at a rate slightly higher than that of the Euro-American population. Jensen, in contrast to the eugenicists of the early twentieth century, argued in his 1969 article that the differential birthrate was favoring Negroes and that this was of great national concern.

Jensen was not alone: he was influenced by several other eugenicists, including physicist William Shockley and psychologist Hans Eysenck. Jensen had spent time as a postdoctoral research scientist at the University of London with Eysenck, a strong opponent of environmentalism and a former student of Sir Cyril Burt (author of the original fraudulent twin studies claiming that intelligence was 80 percent genetic). Upon his return from London, Jensen rapidly revised his earlier positions on genetics, intelligence, and race. He now proposed the simplest hypothesis to explain the relationships between these variables, that is, that genetics, not environment, was responsible for differences in racial ability.

Probably the most outspoken scientist on the danger of dysgenesis in this period was the Nobel laureate William Shockley of Stanford University. Shockley revisited the ideology of classical eugenics in his Nobel address, focusing on the danger of the reproduction of inferior genetic strains. In 1966, Shockley appealed to the National Academy of Sciences for a study of the racial basis of the heredity-poverty-crime nexus, predicting that a high percentage of Negro genes would be found in criminals and the poor. Throughout Shockley's campaign on race, he would accuse his critics of wishing to stifle the free investigation of scientific hypotheses. Scientific arguments raised against his positions he called egalitarian dogma and antigenetic. This tactic is still used by modern psychometricians. Jensen had entered into discussions with Shockley in 1967, when the former was a fellow in behavioral sciences on the Stanford campus. In particular, Shockley wished to convince Jensen that genetics and not environment was the most salient variable in explaining social differences in intelligence.

Jensen began his 1969 *Harvard Educational Review* article with the ominous statement that "compensatory education has been tried and it has apparently failed."[3] He argued that intelligence was determined largely by genetics and that robust intelligence differences had been observed among races. He contended not only that African Americans showed the lowest intelligence scores (15 IQ points, or 1 standard deviation, below those of whites) but also that African Americans and other lower IQ types had greater rates of reproduction. He warned that the United States was in danger of intellectual dysgenesis owing to social welfare programs fostering a high African American reproductive rate. He suggested that early intervention programs, since they could have no permanent impact on the genetically limited intelligence of blacks and other racial minorities, were useless.

Jensen's article, the longest in the history of the journal, generated a flood of media reports, rating lead stories in *Newsweek* (circulation 2,150,000), *Life* (7,400,000), *Time* (3,800,000), and *U.S. News and World Report* (1,625,000). Jensen's article, in fact, garnered more coverage than the discovery of DNA as the heredity material or the Nobel prizes won for virology and medicine in 1954. However, most noticeably, it received more attention than Olive Walker's contradictory study of the Windsor Hills Elementary School in Los Angeles. On October 12, 1969, a *Los Angeles Times* headline read "Black School Highest in IQ—Is

Affluence the Reason?" referring to Walker's demonstration that the G.I. Bill had enabled the parents of the elementary school children to get a college education and that this education had in turn influenced the children's excellent reading and IQ test scores.[4]

Jensenism at Gale Force

The implications of Jensen's report were felt in Washington. Although Richard Nixon had already decided to slow the rate of desegregation in the South, Jensen's article led Nixon to discuss upcoming desegregation cases with Chief Justice Warren Burger, in violation of the principle of the separation of powers (this according to Watergate conspirator John Ehrlichman). In 1970, the winds of Jensenism were, according to Senator Daniel Patrick Moynihan, gusting through the capital at "gale force."[5] In the September 1971 issue of the *Atlantic Monthly*, Richard Herrnstein published an article, simply titled "IQ," that reiterated the core points of psychometric logic. Herrnstein stated that in a society where status is decided by merit, the most important factor is general intelligence. Since IQ (according to Jensen) was determined almost entirely by the genes, American racial stratification was not the result of social discrimination but rather the product of genetically determined cognitive potential.

Within hours of the appearance of the *Atlantic Monthly* article, Eric Sevareid of CBS commented on the air that scientific research had revealed that some people were genetically less educable than others and that the time had come for an agonizing reappraisal of government educational policies. In 1973 Herrnstein would expand his IQ article into a book titled *I.Q. in the Meritocracy*. This work would spawn its own wave of popular articles, including one in the influential magazine *Fortune* in May of the same year.

The scientific response to Jensen, Shockley, and Herrnstein in the 1970s was extensive. Anthropologist Ashley Montagu edited a volume called *Race and IQ* (1975), which contained chapters by some of the most accomplished anthropologists, psychologists, and geneticists of the period, including C. Loring Brace, William F. Bodmer, Theodosius Dobzhansky, Jerome Kagan, Stephen Jay Gould, and Richard Lewontin. Behavioral geneticist Jerry Hirsch and his colleagues also made, and continue to make, important contributions to the scientific critique of Jensenism. These critiques focused on the core logical flaws of the psychometric program, in particular on its reliance on faulty estimates of the heritability of intelligence and the fallacy of the race concept itself. A fifth edition of Montagu's classic *Man's Most Dangerous Myth: The Fallacy of Race* was released in 1974. C. Loring Brace notes in his foreword to the sixth edition that the arguments raised by these scientists seemed to deter psychometric research on race in the 1970s. Nevertheless, psychometric research continues, and the most recent iteration of this program is illustrated by the publication of *The Bell Curve* in 1994.

Eugenics by Any Other Name

The Bell Curve was the expansion of Herrnstein's original theories in his *Atlantic Monthly* piece, with some additional sociological analysis by the conservative Charles R. Murray. Together Herrnstein and Murray resurrected the ideas of the innate cognitive inferiority of African Americans, the declining national IQ (dysgenesis), the lower intellectual quality of immigrants, and the wastefulness of environmental solutions to these problems (Head Start, affirmative action, multiculturalism, and so on). The book was Herrnstein's last attempt to champion his meritocracy argument, for he died shortly after the book's publication. Like Jensen's work two decades earlier, the results reported in *The Bell Curve* were uncritically accepted by elements of the popular press. *Newsweek* ran a cover story on the book, *Time* reported on it, and Diane Sawyer interviewed Charles Murray for ABC's *Prime Time Live*. In addition Murray made a number of other television and radio talk show appearances. Fifty-two of Murray's academic supporters published a signed statement in the December 13, 1995, *Wall Street Journal* entitled "Mainstream Science on Intelligence." *The Bell Curve* sold at least five hundred thousand copies within its first six months. Despite the impressive sales, its impact was hard to gage, although we can gain some insight by reading the statement of Dr. Francis Lawrence, president of Rutgers University: "Let's look at the SATs. The average SAT for African Americans is 750. Do we set standards in the future so that we don't admit anybody with the national test? Or do we deal with a disadvantaged population that doesn't have the genetic hereditary background to have a higher average?"[6]

Although Lawrence later claimed that his original comments had been misunderstood, they could be interpreted in only one way, that is, as saying that African Americans were intellectually inferior to Euro-Americans. In the fall of 1999, an official at the Educational Testing Service suggested that African Americans should receive additional points on their SAT scores to equalize their mean with that of Euro-Americans. In a sense, African Americans would receive points simply for being African American. This procedure reinforces the idea that African Americans are innately less capable of performing well on standardized tests.

In this controversy few people pointed out that the SAT is generally a weak predictor of the eventual success of any college student, let alone that of African Americans. The SAT is known to underpredict African American graduation rates from college and to predict only about 15 percent of the variance in freshman grade point averages. Therefore, we must question the correlation between SAT scores and intelligence.

The Argument for the Link between Race and IQ

Herrnstein and Murray base their claim for a genetic basis to the black-white differential in IQ on the following evidence: First, the National Longitudinal Survey of Youth 1979 found a 1.21 standard deviation difference favoring whites over blacks on the Armed Forces Qualifying Test (AFQT). The sample included 6,502 whites and 3,022 blacks (see figure 10.1). Herrnstein and Murray claim that this IQ difference is evidence of a genetic difference between whites and blacks at loci that influence IQ intelligence. They further support this claim by suggesting that this differential has been robust over the record of mental testing in the twentieth century.[7] In addition, Herrnstein and Murray present data suggesting that the black-white differential does not disappear as one alters parental socioeconomic status.[8] It is this sort of evidence that they marshal to substantiate the claim of an underlying genetic component for intelligence that differs between blacks and whites.

For a moment, let us leave aside any questions about the quality of these data. Taking the data at face value, the naive observer might think that they support Herrnstein and Murray's argument. However, to a professional geneticist their argument is utterly flawed. It is impossible from phenotypic data alone to apportion genetic and environmental sources of causality. Nor is the "robustness" of the data over the twentieth century any indication of genetic causality because the data do not take environmental effects into consideration. For example, pine trees growing on windy coasts show a characteristic twisted body form. As long as these populations live in this environment, the body form will persist. Take their seeds and plant them in a nonwindy environment and the "standard" nontwisted growth form is observed.

Consider another example, this time from the insect world. Fruit flies maintained in the laboratory at 20° C live longer than genetically identical flies held at 25° C. The greater longevity persists over many generations with little genetic change resulting in the stocks.[9] The data summarized in figure 10.1 are taken from only seventy-two years, or about two and one-half human generations. This is not enough time for any substantial

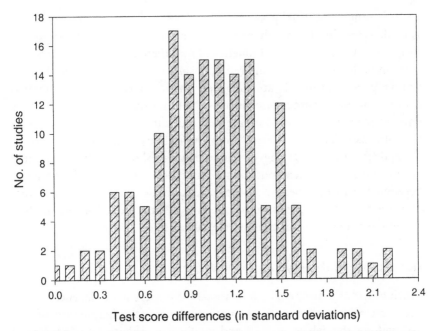

Figure 10.1. Reputed distribution of IQ test differences for blacks and whites over the twentieth century. After R. J. Herrnstein and C. Murray, *The Bell Curve: Intelligence and Class Structure in American Life* (New York: Free Press, 1994), 277.

change in a population of large size, unless it is subject to catastrophic mortality or infertility, which is demonstrably not the case here. It is very important to note that the mean IQ of both Euro-Americans and African Americans has increased by about 15 points (or 1 standard deviation) in the last thirty years (the Flynn effect).[10] Recent data suggest that the rate of increase for average IQ scores became even greater in the 1980s (3.0 points per decade between 1932 and 1978, and 3.5 points between 1978 and 1989).[11] Herrnstein and Murray dismiss this fact by arguing that the increase is more likely a result of better testing than of a real change in mean IQ.[12] Either way, this change indicates an environmental effect, since no significant genetic change could have occurred in this period. This magnitude of environmental effect demolishes the case for the use of raw IQ data as a source of genetic information.

Such inconsistencies also demonstrate that the psychometricians have only an amateurish grasp of evolutionary genetics. Numerous errors flow from this lack of scientific perspective. The vast majority of Herrnstein and Murray's evidence is based on phenotypic information; that is, the reputed difference between races is based on some indirect measure of cognitive function, usually a standardized test. The reliance on such tests is made worse by the fact that they have not been conclusively shown to properly measure intellectual function. From these tests, psychometricians infer an underlying genetic difference, despite the fact that standard quantitative genetic protocols are premised on the extensively corroborated demonstration that procedures such as theirs are scientifically invalid. The flaws in their research program are best illustrated in their obsession with the false association between "group heritability" and the necessity of racial differences in IQ. They presume they can show that IQ is inherited genetically and that there must be substantial genetically based differences between the races. Their focus on genetic predestination of intellectual ability is thus their rationale for supporting the status quo because, in their rather limited view, genes cannot be easily altered.

Heritability in the Psychometric Worldview

Much of the argument concerning racial differences in intelligence has focused on an inappropriate assumption, that is, the heritability of IQ. Throughout the debate on race and intelligence, the methodologies used to establish heritability have been fraught with error and fraud. Psychometricians often ignore basic difficulties in estimating quantitative genetic formulas for variation. The fact that IQ test scores have a continuous distribution indicates that whatever cognitive functions are related to these tests must be influenced by many genetic and environmental factors. The formal expression for heritability in the broad sense is simply the ratio of variance in the character due to genetic sources, over all sources of variance. Direct estimates of heritability in laboratory studies can be tedious. They require rigorous control of confounding environmental factors and careful measurements of the phenotype in question.

Consider the equation for V_p, the variance in the phenotype:

$$V_p = V_g + V_e + V_{g \times e} + Cov(g,e) + V_{error},$$

where V_g = variance of genetic origin, V_e = variance of environmental origin, $V_{g \times e}$ = variance due to gene × environment interaction, $Cov(g,e)$ = the covariance of genes and environment, and V_{error} = variance due to errors in measurement. This equation illustrates that one cannot infer that a phenotypic difference between two groups automatically indicates a genetic difference. Under laboratory conditions we can control the environment such that we can eliminate the third and fourth terms of the equation. For example, if we measure the longevity of fruit flies from two different populations and hold all environmental conditions the same for both groups, then we can safely assume that the third and fourth terms are close to zero. This leaves

$$V_p = V_g + V_e + V_{error}.$$

If we have carefully measured the longevity phenotype, then we can assume that the difference between the two populations is indeed due to genetic sources. However, there is an additional caveat: before we can make these measurements we must rear the flies under identical conditions for at least two generations because complex phenotypes are strongly influenced by maternal environmental effects. The environmental conditions experienced during development can influence the expression of genes in the adult. It should be clear that none of the rigorous controls that are required to identify genetic effects in the laboratory exist under the conditions in which attempts to measure human IQ have been made.

Psychometricians emphasize the heritability of intelligence. But the particular estimate of the heritability of intelligence, however defined, has little to do with the question of cognitive differences between races because the estimates used to calculate the heritability of intelligence result from studies of close relatives. We already know that most of the genetic variability in the human species is at the level of individuals or families. But family-level variation does not therefore translate directly into racial variation. Data from an experiment in my laboratory examining the effect of a known genetic substitution on the complex trait of longevity revealed significant variation in families within populations but no significant variation between the populations. That is, both populations had family genetic backgrounds that responded differentially to the genetic substitution when measured under rigorously controlled environmental conditions. This is another way of saying that if genes do influence intelligence, then we should expect that all races will have families that run the range of the genetic variability for intelligence. Thus, given the large genetic overlap of human populations, our expectation should be that there is no significant racial difference in intelligence or other behavioral traits.

To this prediction the racists will howl, How then do you explain the persistent IQ differential reported by twentieth-century studies? The answer is elementary; let us look at the conditions under which the tests were given. Do they really adhere to the requirements of a valid test of genetic differentiation? Absolutely not.

The problems of the psychometric program do not improve when it attempts to look at specific "genetic" systems reputedly associated with intelligence. After all, Arthur Jensen even admitted that there should be many genes that impact the expression of intelligence, precisely because it is a polygenic trait. It is significant that the psychometricians have been unable to properly define the physiological traits that are purportedly responsible for intelligence and that are differentiated among the racial groups. This lack of precision makes attempts at localizing the genes involved very difficult. For example, a recent study of genes associated with IQ in Chile found that the B allele of the A, B, O locus was correlated positively with IQ score in boys but negatively in girls.[13] The authors of the study were convinced that the large size of the statistical correlation guaranteed that they had uncovered a real genetic contribution made by this allele to intelligence. If these correlations were generally real, then we would expect to find a large variation within so-called racial groups for the genetic contribution to intelligence caused by the blood group alleles.

Table 10.1, which shows the range of frequencies for the B allele in Africa, Asia, and Europe, indicates that there is a large overlap of the frequencies of the B allele among the populations of these regions. Therefore, if the impact of the B allele were independent of genetic background and environment, we would expect its genetic contribution to intelligence to be found in all races. Particularly interesting is the frequency of the B allele in purportedly genetically superior races. For example, in Europe, the English, Germans, and Swedes have very low frequencies of the B allele, 0.061, 0.08, and 0.07, respectively; whereas the "inferior" Slavs have high frequencies (Czechs and Yugoslavs at 0.222 and 0.143, respectively). In the United States, southern whites were shown to have a frequency of 0.05, and blacks a frequency of 0.129.[14] Thus this example of a genetic locus purportedly associated with higher intelligence contradicts the historical claims of the genetic inferiority of Slavs and Negroes relative to Nordics and American whites. For the psychometricians to maintain their support of genetic differences contributing to the intelligence variation among races, they would have to ignore results such as this. Alternatively, they could argue that complex genetic interactions, such as epistasis and gene × environment effects could account for the apparent discrepancies between the Chilean study and their historical quantifications of racial intelligence. However, if they admitted the latter point, they would also have to concede that these factors might make intractable the process of assigning causality for variation in intelligence to specific genetic loci for specific populations.

In the end, the data that the psychometricians rely on to demonstrate racial difference in intelligence are simply the racial differences we already observe. William Shockley, for example, proposed that skin color

Table 10.1. Range of B Allele Frequencies for Various Populations

Region	Range of allele frequencies	Median frequency	No. of populations sampled	Range of sample sizes
Africa	0.019–0.272	0.123	21	95–771
Asia	0.058–0.337	0.193	47	108–65,743
Europe	0.032–0.222	0.071	27	104–100,000

SOURCE: These data are compiled from A. K. Roychoudhury and M. Nei, *Human Polymorphic Genes: World Distribution* (Oxford: Oxford University Press, 1988), table 141.

was the metric by which we could measure intelligence—this despite no established physiological link between the loci that determine skin pigmentation and those that determine any aspect of mental functioning. Nor have the psychometricians been able to advance any credible evolutionary genetic mechanism to explain the origin of these consistent racial differences. We know that genetic differences among populations are created by the combined action of natural selection and genetic drift. The selection clines involved in producing human genetic variation differ independently from one another. There is no reason to suppose that these should have produced intellectual inferiority only in sub-Saharan Africans. Genetic drift cannot be invoked either, since drift events are random, and thus allelic variation related to intelligence that results from drift should also be scattered throughout human populations, as the case of the B allele illustrates. Thus, to explain the consistency of inferior IQs in sub-Saharan Africans, one would have to suppose some form of natural selection that was operating only on these populations. J. Philippe Rushton attempts to accomplish this by utilizing an r- and K-selection scenario to explain the life history features of the three major races. Briefly, he argues that the human races fall on the r- and K-continuum. The theory of r- and K-selection was devised in the late 1970s to address why some species had short lives and reproduced rapidly, while others had long lives and reproduced slowly. Examples of organisms on different ends of the scale would be weeds, which grow rapidly, but invest little in their bodily structures (r-selected) and trees, which grow slowly and invest large amounts in their structures (K-selected). According to Rushton's view of the human races, Negroids are considered "weeds," with high investments in reproduction, and thus less to invest in bodily structures such as brain mass, thereby having lower intelligence. Alternatively, Caucasoids and Asians are more "treelike," with high investments in brain mass and thus greater intellect, and lower inputs to reproduction. I have examined his scenario and have argued that he fails.[15] This failure results from both an improper use of life history theory and a flawed analysis of the available data.

The psychometric argument gets weaker when we examine the genetic variability within the human species, particularly the greater within-group than between-group variation. On solely genetic grounds one would expect these racial groups to be indistinguishable for a complex behavioral trait like intelligence. Indeed, even if intelligence were highly heritable, the establishment of differences in intellectual capacity among family groups within a supposed race would not mean that there would be differences among races. The argument for consistent genetic differentiation for IQ among races suffers from all of the points that I have raised. Each alone is a fatal error, and when taken together they invalidate the racist program of Shockley and his co-conspirators.

The Ubiquity of Gene × Environment Interactions

The most elementary requirement for a test of genetic differences between phenotypes is that they have been reared in and tested under the same environmental conditions. Meeting this requirement is not as easy as it seems, even with relatively simple traits. It is clear that no such equality of environmental conditions has existed or now exists for the races in the United States. In chapter 5, I discussed how the impact of hookworm infections on children varied depending on income and on soil type. We also know that the environmental conditions for African Americans were horrendous at the turn of the twentieth century. The entire history of the United States shows that the environment for African Americans has been consistently inferior to that for Euro-Americans. By necessity, that fact makes problematic any attempt to apportion genetic causality to any complex phenotype differing between these groups. For example, the scores on the AFQT surveyed by Herrnstein and Murray came from students who attended school systems that are still mostly segregated. Consider the segregation index for African Americans in the following cities (1993 data): Gary, Indiana, 89 percent; Detroit, 88 percent; Chicago, 86 percent; Cleveland, 86 percent; Milwaukee, 83 percent, Buffalo, 83 percent; St. Louis, 81 percent; Philadelphia, 80 percent; Cincinnati, 79 percent, and Birmingham, 79 percent. In the period from 1968 to 1992, between 67 and 76.6 percent of African American children attended schools with more than 50 percent minority enrollment. Numerous contemporary studies have shown that these segregated school districts are disadvantaged relative to white districts.[16] Historical segregation and the unequal access to educational resources it caused are crucial in interpreting the results of standardized tests. Studies have shown that students' scores on the mathematics portion of the SAT are strongly correlated with the number of mathematics courses they have taken in high school. Poor minority districts often have difficulty in attracting qualified mathematics and science teachers. Thus students in these areas probably, if not certainly, received lower quality instruction.

There is also a growing awareness that minority populations are often forced to live in environments contaminated by toxic materials. We know that some of these materials affect the cognitive development of children who are exposed to them. For example, inner city children are often exposed to high levels of lead, which causes mental retardation and impaired growth; and children of farm laborers are exposed to high levels of pesticides with known neurotoxic impacts. As one might expect, racial composition is the best predictor of the location of hazardous waste facilities in the United States. This is true even when socioeconomic class is included in the statistical analysis. Thus, poor whites receive more toxic exposure than wealthy whites but less than poor racial minorities. This stratification means that many minority communities are literally on a toxic treadmill, with jobs in their communities being located in or near hazardous waste–producing industries or disposal sites. The problem of toxic materials and race in the United States is chilling in scope. Consider the following facts:

- The nation's largest landfill, which receives toxic materials from forty-five states and several foreign countries, is located in Sumter County, South Carolina, which is predominantly made up of poor African Americans.
- The predominantly African American and Hispanic Southside of Chicago has the greatest concentration of hazardous wastes in the nation.
- In Houston, Texas, six of the eight municipal dumps, and all five landfills, are located in predominantly African American neighborhoods.
- In West Dallas, Texas, one African American neighborhood's children suffered irreversible brain damage from exposure to lead at a nearby smelter. The residents won a $20 million out-of-court settlement.
- Pesticide exposure among Hispanic farm workers causes more than three hundred thousand illnesses a year. A large percentage of these workers are children and women of childbearing age.

Table 10.2. Toxic Waste Site Locations and Race in Selected Cities

City	No. of sites	African American population (%)
Memphis, Tenn.	173	43.3
St. Louis, Mo.	160	27.5
Houston, Tex.	152	23.6
Cleveland, Ohio	106	23.7
Chicago, Ill.	103	37.2
Atlanta, Ga.	94	46.1

SOURCE: C. Lee, "Toxic Waste and Race in the United States," in *Race and the Incidence of Environmental Hazards*, ed. B. Bryant and P. Mohai (Boulder, Colo.: Westview Press, 1992), table 1.

There are more instances. Particularly illuminating is the percentage of African American population in the six cities that lead the list of cities with the highest number of hazardous waste sites (see table 10.2). Presently, 53 percent of the toxic waste sites located within one mile of public housing projects are located in communities that are greater than 75 percent minority (a total of 2,628 sites).[17] These data are immediately relevant to the issue of innate racial differences in intelligence. A large sector of the African American and other minority populations are currently forced to live in poisonous environments. This is also reflected in the lower life expectancies of African Americans, as illustrated by their age-specific mortality (chapter 11). To be valid, any test of the genetic hypotheses of racial differences in intelligence must at least equalize the physical and educational environment of Euro-Americans and African Americans. The problem is not only that this experiment cannot be performed under the existing political circumstances but also that the proponents of the link between race and IQ do not argue that the experiment should be performed to test their hypotheses.

CHAPTER 11

The Race and Disease Fallacy

Even my most adept students often suggest that disease predisposition differs between human races. Indeed the biomedical research community has recently begun to see the importance of including races other than whites in their studies of specific diseases. In describing studies that focus on populations other the Euro-American males, the National Institute on Aging and the National Cancer Institute now use the term "special populations." The problem is that the populations they define as "special" include "minorities, women, and the disabled." In reality, all groups are minorities at the level of genetic variation, including Euro-American males; and women are the majority sex. The use of "disabled" is even more problematic, since any person can go from able to disabled in the time it takes a car to crash. The lack of precise definition here is crucial. Research programs that improperly define the subjects they wish to study will by necessity provide compromised answers.

Some pharmaceutical companies are now advertising the significance of race in the action of their products. SmithKline Beecham included the following message concerning the benefit of calcium supplements on bottles of TUMS E-X released in 1999: "IMPORTANT INFORMATION ON OSTEOPOROSIS: Research shows that certain ethnic, age, and other groups are at higher risk for developing osteoporosis, including Caucasian and Asian teen and young adult women, menopausal women, older persons, and those with a family history of fragile bones. A balanced diet with enough calcium and regular exercise throughout life will help you build and maintain healthy bones and may reduce your risk of developing osteoporosis."

This message suffers from the common misconceptions about human biological variation and disease. The first is the confusion of ethnic groups and races. Ethnicity is defined culturally, whereas the concept of race relies on presumed biological variation, which in this case is being used to understand the predisposition for disease. The second problem is that all humans have some risk for osteoporosis, and the recommendations made concerning a balanced diet and exercise are useful for everyone. Thus, there was no need to make the statement that certain ethnic groups are at higher risk, when all groups have some risk. The problem is that many biomedical researchers and clinicians are still working under the yoke of the biological race concept. Hence, they see all biological differences between and within populations as potentially due to racial genetic composition.

"The Health of the Negro"

Western medical interest in Africans began in the era of chattel slavery. One of the first well-known medical examinations given to slaves during warehousing and the Middle Passage was the search for Winterbottom's sign. First described by a British physician, the sign is a swelling of the lymph nodes associated with acute infections of trypanosomiasis (African sleeping sickness). Slaves found with Winterbottom's sign during the Middle Passage were dumped overboard because they would not survive the journey. Obviously, as with other beasts of burden, healthy slaves were more valuable than sick ones.

Information about either predisposition or resistance to disease would have been useful to slaveholders. One of the first observations of disease resistance concerned yellow fever. As we have seen, slaves in the Caribbean showed greater resistance to yellow fever than their masters. Darwin also commented in *The Descent of Man* that explorers in the tropics succumbed to diseases more often than did the natives of the region. Conversely, one of the largest sources of mortality for the native Indian tribes of the Western Hemisphere was the introduction of new pathogens originating in the Eastern Hemisphere. Epidemics of these pathogens decimated native populations before they could acquire resistance.

Neither the meager medical care provided to slaves by the master nor the medical knowledge available in their own culture had any significant impact on reducing slave mortality. Calculations indicate that the mortality of slaves throughout the New World was an order of magnitude higher than that suffered by the Europeans. The slaveholders' medical response to the misery of slavery was to invent new diseases, such as draptemania (the insane desire to run away) or dysesthesia (a disease that caused slaves to work in a haphazard or lazy manner). The cure for these was to treat the slaves in a kind and paternal way; if this approach failed, there was always the lash.

The abolition of slavery did not readily improve African American health. Initially, most freed slaves were without work or land, and they had no training in health precautions. There were very few medical practitioners or facilities in the African American community. We have already seen how the biological theory of the period continued to see Negroes as a degenerate race. Social Darwinians felt that the Negro was doomed to extinction on the North American continent. The low birthrate and high mortality rate exhibited by African Americans in this period supported this prediction. In "Darwinian" competition, African Americans were simply losing.

The public health community's interest in the Negro was primarily due to the desire to protect whites from diseases that seemed to originate or permanently reside in the Negro community. Of particular concern were communicable diseases, such as syphilis or tuberculosis, that might spread to whites. Secondarily the public health community was concerned with the economic losses that resulted from untreated diseases among Negroes in the South. The Tuskegee syphilis project of 1932–1972 (discussed later in this chapter) is one of the greatest tragedies in the history of American medicine and public policy. This project began because physicians felt that race was a meaningful biological category with the ability to accurately predict disease susceptibility. Medical practitioners then, as now, had an incomplete grasp of evolutionary theory. What they did understand came mostly from the eugenicists. Therefore, heritable genetic diseases were a major focus of concern. The early study of sickle cell disease was tainted by fear of the spread of the trait into whites. Sickle cell disease was hypothesized as a dominant genetic trait specifically for use as a propaganda tactic to help prevent race mixing. In contrast, no such racial stigma was attached to genetic diseases found predominantly in northern European populations.

The Great Racial Health Gap

The racial disparity in health and survival persisted throughout the twentieth century. In his February 21, 1998, radio address, President William Jefferson Clinton committed the nation to eliminating the disparities in six areas of health status exhibited by racial and ethnic minority populations by the year 2010. The working hypothesis of many within the medical community concerning racial differentials in disease incidence is still heavily genetic. In fact, the traditional medical research literature assumes racial genetic difference as if it were a fait accompli in explaining disease prevalence (as shall be demonstrated in this chapter).[1] This assumption is made without consideration of what is known about the nature of human genetic variability. For example, the higher African American age-adjusted mortality in twenty-two of twenty-four mortality categories is not considered problematic by this school of thought (age-adjusted mortality refers to standardizing age distributions between sample populations). What is the probability that a combination of natural selection and genetic drift produced such a one-sided asymmetry in mortality, especially when those with higher mortalities are hybrids of three types of populations? We should remember that African Americans are hybrids of Africans, Asians, and Europeans. However, which populations make up the hybrid is irrelevant. What is relevant is that the populations are hybridized and thus should have heterozygote advantage. African Americans would have the lowest disease rates if genetics were indeed the key. The fact that African Americans show higher rates of biological mortality due to tuberculosis, diabetes, hypertensive heart disease, ischemic heart disease, hypertension, cerebrovascular diseases, intracerebral hemorrhage, thrombosis, pneumonia, chronic obstructive pulmonary disease, ulcers, chronic glomerulonephritis, and renal failure cannot be explained by different gene frequencies. In fact, in any other field, an extremely powerful evidentiary base would have to be advanced and supported to make this kind of claim. Thus far, concerning the greater mortality of African Americans, this has not been the case.

The difference between African American and Euro-American mortality patterns in the twentieth century is staggering. Figure 11.1 shows the ratio of the age-specific mortality from all sources for these two populations in 1963, 1980, and 1996. The figure shows that African American infant mortality is between 1.8 and 2.5 times higher than that of Euro-Americans, and the age-specific mortality of African Americans always exceeds that of Euro-Americans, except at the oldest ages (greater than 85 years). The robustness of these patterns, similar to that discussed earlier for the IQ differential, has led some to suggest a solely genetic explanation. The question still remains, however, whether either the theory or the experimentation supports such assertions. We shall see that, just as for IQ, the answer is no.

The Racial Distribution of Cancer

The very word "cancer" strikes fear into the hearts of most people. Despite the recent advances in detection and treatment, most still view the disease as a death sentence. Even more frightening is the concept that because of one's ethnicity or "race," one might somehow be singled out to acquire this disease. This fear is made worse by medical practitioners who inadvertently reinforce racial stereotypes in their approach to diagnosis and treatment. These stereotypes have fostered the belief in many minority

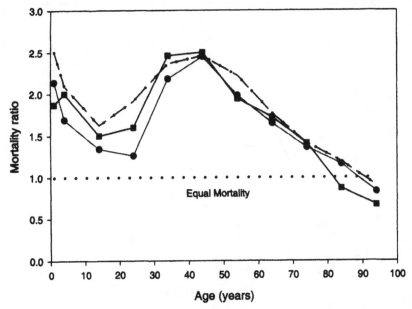

Figure 11.1. Ratio of black/white age-specific mortality rates (per 100,000) from all recorded causes of death in 1963, 1980, and 1996: —•— = 1996; —●— = 1980; and —■— = 1963. The data from the three periods show high concordance, which suggests that the black/white mortality ratio has not changed since 1963. U.S. Census Bureau, *Statistical Abstract of the United States* (Washington, D.C.: GPO, 1999), 96.

communities that they are somehow genetically predestined to die from this disease.

For example, consider the black/white ratio of reported deaths from all cancers in 1995 (see figure 11.2).[2] The ratio for total cancer deaths shows the same general pattern as the ratio for total mortality plotted in figure 11.1. There are, however, many forms of cancer, and each form is influenced by different genetic systems. This suggests that it is unreasonable to expect that all forms of cancer should have higher incidence in one race. In fact, for the twenty-four cancer types examined in a 1985 study, African Americans exhibited higher incidence rates than Euro-Americans in fifteen.[3]

B. A. Miller and coworkers' 1996 examination of cancer incidence rates sorted by cancer site and by racial or ethnic group is also revealing in this regard (see table 11.1). These data indicate that the rank of a population varies for each type of cancer and that even closely related populations may have very different incidence rates for specific types of cancer. Table 11.2 ranks the populations listed in table 11.1 in order of the

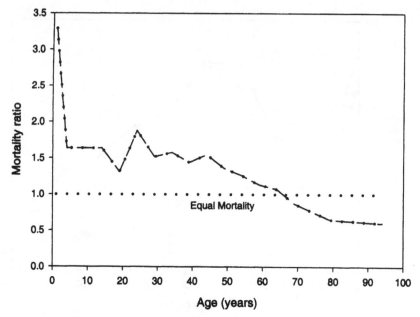

Figure 11.2 Ratio of black/white age-specific mortality rates from all cancers in 1995 — · — · — · —, • • • • • = equal mortality rate for whites and blacks. U.S. Census Bureau, *Statistical Abstract of the United States* (Washington, D.C.: GPO, 1998), 102.

frequency of stomach cancer versus that of lung and bronchial cancers. In these populations, there is no apparent evidence for clustering of cancer incidence based on genetic relatedness. The lack of clustering is due in part to the strong influence on cancer of environmental factors such as diet, water quality, air quality, and smoking.

In addition, the way the data are reported in this study makes analysis of any genetic hypothesis difficult. In particular, Asian subpopulations are reported by country of origin, whereas the groups labeled black, Hispanic, and white are not so differentiated. Even though blacks, as we have seen, can have quite different genetic histories, everyone from newly immigrated sub-Saharan Africans to Caribbeans and those Americans historically defined by hypo-descent fall into this category. The Hispanic category is equally problematic, representing once again a linguistic, cultural cluster, not a genetic lineage; Chicanos, Puerto Ricans, and Argentinians are lumped together as Hispanic. Finally, the white category could contain individuals from regions as diverse as Scandinavia, Turkey, and Egypt.

Miller's data for breast and prostate cancer reveal a similar pattern (see table 11.3.). The mean cancer rates in the individual Asian populations are treated in a manner analogous to those of the black, Hispanic, and white categories. As a result, the data suggest that cancer rates in the Asian category are generally much lower than the rates in the other categories. These data could be used to support the idea that these categories are genetically differentiated for cancer predisposition. However, when the individual Asians groups are combined into one large category, these data no longer support this idea.

In addition, there is strong evidence that there are uniquely American environmental cues that play a central role in the onset of cancer. For example, it has been shown that the longer immigrants live in the United States, the more their cancer rates mirror those of other United States groups. Black, Hispanic, and white immigrants have, for the most part, lived in the United States longer than Asians immigrants, who began migrating only relatively recently (particularly to the American mainland). This change over time is similar to the observation of changes in immigrant body form made by Franz Boas in the 1920s. In any event, the fact that these populations cannot be shown to share similar migratory patterns or other common environmental factors vitiates any genetic analysis of either their cancer predisposition or their IQ potential.

Even within the larger racial categories, such as in the populations collectively called "Asians," the data in table 11.3 show that relatively closely related populations may have very different incidence rates for specific cancers. The variance in breast and prostate cancer rates within Asian populations is considerable. Japanese and Korean populations are genetically closely related, yet show very large differences in the incidence of these specific cancers.

The lack of a pattern is also apparent in data from international populations. If we examine international cancer incidence and mortality data by race, no appreciable pattern emerges. Table 11.4 compares 1987 WHO mortality data for breast cancer in Caribbean and Scandinavian countries. Although the Caribbean populations are an admixture of African, American Indian, and European populations, it is clear that the Caribbean and Scandinavian populations overlap in their breast cancer mortality rates. The incidence data for the United States were three to four times higher than the international figures in the 1980s. This result immediately invalidates the racial explanation and instead raises the question whether there is a uniquely American environmental factor that may be accounting for this difference.

One such mechanism that has been suggested is the American diet; in particular, diets high in fat have consistently been associated with greater cancer incidence. The observation that within a few generations immigrant

Table 11.1. Cancer Incidence (per 100,000) at Selected Sites for Racial and Ethnic Groups, Age Adjusted to 1970 U.S. Standards

Racial or ethnic group	Stomach	Colon or rectum	Lung and bronchus	All sites
Native Alaskans	13.6	**73.5**	65.8	360
American Indians, N. Mex.	0	*16.95*	*7.2*	*188*
Chinese	9.5	39.2	38.7	247.5
Filipinos	*6.9*	*28.15*	35.0	249
Hawaiians	16.75	36.45	**66**	330.5
Japanese	**22.9**	51.8	*29.1*	281.5
Koreans	**34.0**	26.8	34.6	*223*
Vietnamese	25.8	28.8	51.05	299.5
Hispanics	15.3	31.5	*30.6*	281
Blacks	12.75	**53.1**	**80.6**	**443**
Whites	*9.6*	48.4	61.5	**417.5**

SOURCE: B. A. Miller et al., eds., *Racial/Ethnic Patterns of Cancer in the United States, 1988–1992* (Bethesda, Md.: National Cancer Institute, 1996), 22, 41, 73, 113.
NOTE: The incidence rates in the table have been adjusted to reflect the 1970 U.S. population age structure. This adjustment is necessary because it is not possible to directly compare populations if they have different age-class distributions. Values in italics are the three lowest rates for that category, and values in boldface type are the highest values for that category.

Table 11.2. Rank Order of Cancer Incidence (per 100,000) for Various Racial and Ethnic Groups

Stomach	Incidence	Lung and bronchus	Incidence
American Indians	0	American Indians	7.2
Filipinos	6.9	Japanese	29.1
Chinese	9.5	Hispanics	30.6
Whites	9.6	Koreans	34.6
Blacks	12.75	Filipinos	35.0
Native Alaskans	13.6	Chinese	38.7
Hispanics	15.3	Vietnamese	51.05
Hawaiians	16.75	Whites	61.5
Japanese	22.9	Native Alaskans	65.8
Vietnamese	25.8	Hawaiians	66
Koreans	34.0	Blacks	80.6

SOURCE: B. A. Miller et al., eds., *Racial/Ethnic Patterns of Cancer in the United States, 1988–1992* (Bethesda, Md.: National Cancer Institute, 1996), 22, 41, 73, 113.

Table 11.3. Cancer of Breast and Prostate Incidence (per 100,000) for Racial and Ethnic Groups, Age Adjusted to 1970 U.S. Standards

Racial or ethnic group	Breast	Prostate
Native Alaskans	78.9	*46.1*
American Indians, N. Mex.	*31.6*	52.5
Chinese	55.0	46
Filipinos	73.1	69.8
Hawaiians	**105.6**	57.2
Japanese	82.3	88
Koreans	*28.5*	*24.2*
Vietnamese	*37.5*	*40.0*
Hispanics	69.8	**89.0**
Blacks	**95.4**	**180.6**
Whites	**115.7**	**137.6**

SOURCE: B. A. Miller et al., eds., *Racial/Ethnic Patterns of Cancer in the United States, 1988–1992* (Bethesda, Md.: National Cancer Institute, 1996), 81, 109.
NOTE: The values in italics and boldface type specify the three lowest rates and highest rates, respectively.

Table 11.4. Age-Adjusted Mortality (per 100,000) from Breast Cancer in Selected Caribbean and Scandinavian Countries

	Mortality
Caribbean countries	
Bahamas	25.0
Barbados	22.5
Cuba	**14.3**
Dominica	*39.1*
Martinique	22.3
Puerto Rico	**10.5**
St. Vincent	*28.8*
Trinidad	**16.6**
Scandinavian countries	
Denmark	27.6
Norway	18.2
Sweden	18.1
Finland	16.7

SOURCE: *World Health Statistics Annual*, Geneva, WHO, reprinted in A. P. Polednak, *Racial and Ethnic Differences in Disease* (New York: Oxford University Press, 1989), 70–71.
NOTE: Caribbean figures in italics are higher than the highest Scandinavian rate; and Caribbean figures in boldface type are lower than the lowest Scandinavian rate.

cancer incidence patterns begin to converge on American cancer incidence patterns has been linked to changes in diet as immigrants become "Americanized." This factor may be particularly relevant for Asian populations, whose diets traditionally included greater proportions of fish. Compared to beef and pork, fish contains much higher percentages of beneficial fatty acids.

Genetic Polymorphism and Cancer

If it is true, as has been recently suggested, that only about 5 percent of the differential in racial cancer incidence is genetic, then it would be hard to explain, by genetic polymorphism alone, the large incidence and mortality rate differences between African American and Euro-American populations. However, genetic polymorphisms have been found that are consistent with a small genetically determined component to the observed rates differences. For example, differences in lung cancer rates between whites and blacks might be explained by allele frequencies for the enzyme NAT2, which deactivates carcinogenic aromatic amine groups. Enzymes have characteristic and quantifiable rates of action, and NAT2 has a slower form called NAT-s. The frequency of this slower form differs for African Americans and Euro-Americans: the allele for the slower form is found at 30–40 percent in African American populations versus 50–60 percent in Euro-American populations. The impact of this locus on the production of a cancer phenotype seems entirely dependent on an individual's rate of smoking; that is, the enzyme seems to have a protective impact only with lower exposure to tobacco smoke, indicating its action is strongly influenced by behavioral or environmental variables. Other polymorphisms have been reputedly associated with racial variation in general cancer incidence, for example, the h-ras, 5–alpha-reductase, and NAD(P)H quinone oxidoreductase genes.

There are problems in many cancer studies that suggest that they do not establish causality between the genetic variants and the particular cancer phenotypes. These studies typically identify genetic variation at the molecular level utilizing socially defined groups. These groups also differ in the incidence of the cancer phenotype in question. Often the allelic variants are defined at the molecular level such that individuals carrying these variants are already differentiated between the groups. Thus, this method of identification is circular in that it does not establish causality of the genetic variants a priori.

Recently there has been a great deal of interest in using simple molecular polymorphisms in DNA nucleotide sequences to establish links between race and cancer. For example, it has now been clearly established that certain diseases, such as fragile X syndrome and Huntington's disease, can be caused by different nucleotide sequences produced by triplet mutations. A triplet mutation is the insertion of short tandem repeats of nucleic acids (such as CAG or GCC) into portions of genes coding for a necessary protein. These mutations are related to transposable genetic elements, which are sequences of DNA that code for their own replication. These nucleotide repeats work like a computer virus that makes a cascading series of copies of its own program on a hard drive. As the number of program copies increases, so does the amount of damage it causes, particularly if the virus has infected crucial portions of the drive.

If a computer virus were capable of evolution, the virus found on each infected machine would be slightly different. However, machines that were infected at around the same time would have similar viral codes. In the same way, individuals who are genetically related should share similar short nucleotide repeat alleles. This principle forms the basis for forensic techniques such as DNA fingerprinting. Thus it has been surmised that individuals from the same race should be similar to each other in triplet nucleotide alleles, because they should have inherited them at about the same time. However, finding a common allele within a population does not in itself establish genetic causality for a given disease. It is also necessary to demonstrate that the observed genetic difference does indeed have functional importance. In human studies, demonstrating functional importance requires the analysis of affected and nonaffected members of families carrying the allele of interest.

A recent example of the fallacy of utilizing triplet mutations to connect race and disease comes from studies of prostate cancer. Data demonstrating higher prostate cancer incidence in African Americans stimulated speculation that genetic differences between races could account for the higher incidence. Researchers examining variation in the androgen receptor gene found triplet mutations that varied among individuals. In 1995, a team of southern California researchers examined molecular genetic variation in the androgen gene under the assumption that this variation would vary among racial categories.[4] They determined allele frequency (the number of CAG and GCC nucleotide repeats in the receptor gene) in a population divided into three groups: forty-four African Americans, thirty-nine Asians (Chinese and Japanese), and thirty-nine non-Hispanic whites. Dividing the groups in this way is suspect because an accurate estimation of allele frequencies in any population requires both larger and broader sampling regimes. The study continued by defining allele types and associating them statistically with the incidence of prostate cancer. This association resulted from an earlier study that utilized a population of 57 "white" males affected with prostate cancer. Allele types were defined in the following way: repeat numbers of less than or equal to twenty-two CAG repeats were called "short," and repeat numbers of greater than twenty-two repeats were called "long" (CAG repeat numbers ranged from 9 to 29). These definitions were entirely arbitrary: the dividing line between the two types, twenty-two repeats, is simply the median number of repeats in the Asian population. By applying this measure of long or short to the entire population of the study, the researchers found that the frequencies of short CAG repeats for African Americans, Asians, and non-Hispanic whites were 0.75, 0.49, and 0.62, respectively.

This distribution was determined to be highly significant by statistical analysis. However, if we choose a different criterion (and this possibility was not examined in the paper), we arrive at very different percentages. Choosing the median number of repeats from the African American population (twenty) to define long and short alleles gives frequencies of 0.59, 0.33, and 0.54 (same order as above). Now the allele frequencies of the African American and non-Hispanic white populations are much closer (0.59 and 0.54), raising the question as to whether any significant difference could be supported by this analysis.

An additional problem with the 1995 study is that its authors never give a functional rationale for grouping the alleles into long and short types. The nucleotides in a coding sequence of DNA spell out the sequence of amino acids that will be assembled to construct the protein encoded in that section of DNA, and the amino acid sequence in turn determines the protein's three-dimensional shape. This shape needs to be specific if proteins are to catalyze the specific chemical reactions required for a cell's proper functioning. Thus errors in that three-dimensional shape could be significant. The amino acid glutamine results from the CAG triplet code. If a number of CAG repeats were added to the coding portion of the gene for a particular protein, a series of glutamine amino acids would be inserted into the protein, possibly altering the protein's three-dimensional structure. This structural change would in turn alter the binding properties of the molecule. This explanation was suggested by the authors; however, they did not properly understand its significance. The data they provided actually showed that the number of repeats quantitatively varied within populations (as opposed to discretely, long versus short). Thus the proper analysis of allelic number revealed by this study was not two (short or long), but twenty-one (the number of different triplet repeats observed in the sample).

Another flaw in the 1995 study is its failure to establish the repeat frequencies in men who did not suffer from prostate cancer. This issue is addressed by a 1999 examination of racial variation in repeat number. The new study verified that the populations, as defined by the 1995 study, did indeed differ in repeat number, but the study found that the repeat number also varied in men who did not have prostate cancer. This result illustrates that genetic differences do not necessarily have functional significance. Also in 1999, a French-German research group examined this problem utilizing a better sample design: 105 control individuals, 132 individuals with spontaneous cases of prostate cancer, and 131 individuals from families with a high incidence of prostate cancer (85 affected and 46 not affected). In addition to performing statistical analysis by assuming discrete categories of less than twenty-two and greater than twenty-two repeats, the research group also treated each repeat class statistically as a separate variable, yielding twenty-one alleles. Neither method of analysis found any statistical difference in CAG repeat number between control and affected individuals.[5] Thus, the 1995 study failed precisely because it assumed a correlation among genetic variation, disease, and race. This assumption was not substantiated by the researchers' techniques, yet despite these problems they continued to champion the significance of race in prostate cancer research.

Additional Difficulties of Race in Medicine

Some of the greatest advances in biology from the latter half of the twentieth century occurred in molecular genetics. These advances, however, did not occur without a price, and that price was the rise of molecular reductionism, which resulted in part from the increase in specialization that occurs as the knowledge base grows within the discipline. The problem with specialization is that it almost always entails a lack of communication among allied fields. The lack of communication between evolutionary and molecular geneticists caused failures in both disciplines. Ignoring the significance of molecular mechanisms fostered a naive allegiance to adaptationism in evolutionary biology, and underestimating the significance of evolutionary genetics contributed to misinterpretation of the origin and significance of genetic variation in molecular biology.

For example, cancer research demonstrated the existence of particular allelic variants that seemed to be associated with specific cancer phenotypes. For example, p53, BRCA1, and BRCA2 mutations have been associated with various reproductive cancers. To some researchers, the discovery of these mutations (some revealed only in small populations) in the mid-1990s suggests that long-standing differentials in supposed racial cancer incidence rates might be related to underlying allelic variation among the supposed human races. As we have seen, the combined action of natural selection and genetic drift leads to gene frequency variations in local populations. The problem, of course, is whether the genetic variation for disease incidence can map the socially defined categories of race.

This problem is illustrated by a 1999 study of the supposed connection between BRCA mutations and breast cancer. Studies in the mid-1990s found that Ashkenazi women carrying BRCA mutations had an increased risk of developing breast cancer relative to noncarriers. The 1999 study demonstrates, however, that the BRCA mutations do not seem to be associated with early breast cancer onset in larger outbred populations of women in England.[6] Thus, it is entirely possible that the impact of BRCA mutations among the Ashkenazim results from their particular population history. Genetic diversity in the Ashkenazim was extremely reduced by the Holocaust, and thus the impact of BRCA may be the result of inbreeding depression, the general fitness decline observed in all populations that lack genetic diversity. Since inbreeding depression is not limited to any specific population, we might predict that BRCA1 impacts would be randomly distributed. This again demonstrates that we cannot map its genetic impact onto socially constructed categories of race.

Clinal Selection and Disease Resistance

Evolutionary theory provides a powerful tool indispensable for understanding the genetics behind disease frequency in human populations. There are two types of disease-causing genes relevant to this problem: genes that cause inborn errors of metabolism (normally found at low frequency) and those that play a role in the eventual manifestation of disease (found at higher frequencies). In addition, there are genes responsible for resistance to various diseases. The frequencies of alleles of these genes in specific populations are caused by a combination of clinal selection (selection along environmental gradients) and genetic drift resulting from peculiarities of population history.

Skin color and diseases related to it provide an interesting example of weak selection resulting from clinal geographic variation. A cline is a gradient of continuous variation in a phenotypic or genetic character within a species. There are several possible explanations for the existence of a cline. A cline might result if natural selection favored a slightly different feature at each point along a geographic gradient. A cline could also result from gene flow between two groups previously adapted to different environments. In the Eastern Hemisphere, skin color shows a clear cline related to sunlight intensity. Natural selection established this cline over a long time period. (Populations in the Western Hemisphere, which migrated from the East and arrived in the West between eleven thousand and thirty-five thousand years ago, did not reestablish the gradient.) The distribution of alleles affecting melanin levels in the skin follows this cline, and there is a well-established relationship between population variation and resistance to diseases associated with sunlight exposure. All of the so-called races show high melanin concentrations in the skin in tropical latitudes (Caucasians in India, Australoids in the South Pacific, Asians in Central America and Indochina, and Africans in Africa). Despite this concordance, the incidence of skin melanomas does not directly map skin color. Australoids, who have very high melanin content, show some of the highest rates of melanoma, perhaps as a result of cultural practices that lead to greater sun exposure.

Vitamin D binding protein factor may be important in at least two different diseases, rickets and diabetes milletus. The frequencies for the two major alleles from several populations at the vitamin D binding locus are shown in table 11.5. The lesson is that one cannot assume that variation in specific alleles causally predisposes or protects individual races from a disease. Genetic variation at this locus in Australoids is discordant with their remaining genetic variation. The genetic background might or might not produce a countervailing effect. The fatal flaw of relating disease to race is its failure, in general, to establish mechanistic relationships between gene frequency differentials and differentials in disease prevalence among the populations in question. This can be effectively done for only a few loci of major penetrance. Penetrance is the capacity for a gene to be expressed without regard to genetic background or environmental effects. Thus a gene with weak penetrance may not be expressed for a number of reasons, whereas a gene of major penetrance will almost always be expressed. Yet even these do not map race effectively.

Variations in body stature are another example of clinal selection. Bergman's rule states that a large ratio of surface area to body mass is required for convection, conduction, and radiation of heat. Therefore, according to physiologists, relatively warm climates favor lanky body types, which have more surface area relative to body mass; and cold climates favor more stout body types, which have a lower surface area–body mass ratio. Experiments with humans and many other mammals on shivering as temperatures are lowered and on increase of body temperature as temperatures are increased have verified Bergman's rule in the laboratory. The surface area–body mass ratio is lowest in Eskimos and highest in African groups such as the Nilotes. There is a relationship between this ratio and

Table 11.5. Allele Frequencies at the Vitamin D Binding Locus

	Africans	Europeans	Australoids
Allele 1	0.907–0.963	0.702–0.795	0.824–0.900
Allele 2	0.037–0.093	0.205–0.298	0.100–0.152

SOURCE: A. K. Roychoudhury and M. Nei, *Human Polymorphic Genes: World Distribution* (New York: Oxford University Press, 1988), 150–153.
NOTE: The ranges of alternative allele frequencies are shown for three geographically separated populations.

disease, in that temperature regulation is related to resistance to a number of infectious diseases. For example, heat retention may affect the immune response to respiratory tract diseases, and fever production is involved in resisting viral and bacterial infections.

Temperature and sunlight intensity are not the only climatic factors that produce selection clines. There are other clines that are less obvious. For example, differences in climate may be correlated with general differences in biota, and as we have seen, the fields of tropical medicine and parasitology have long noticed Europeans' lack of resistance to tropical disease organisms. However, one of the most interesting recent discoveries shows clinal variation for disease resistance favoring Europeans over Africans and Asians. European populations have greater frequencies of the molecular variants of the CCR5 protein found on T-cells, which are the primary targets of the HIV virus. Individuals who have the CCR5 mutant allele are resistant to the action of the HIV virus. The mutation is a deletion of thirty-two base pairs in the coding region of the gene that produces the CCR5 protein on the surface of human macrophages (white blood cells involved in the immune response). The loss of these base pairs alters the structure of the protein, most likely interfering with the ability of the HIV virus to bind that particular protein found on the T-cells. The mutant allele (CCR5$^-$) is co-dominant with the more common CCR5$^+$ allele. Thus, homozygous mutant individuals (CCR5$^-$ CCR5$^-$) have greater resistance to the virus than do heterozygous individuals (CCR5$^+$ CCR5$^-$), who in turn have greater resistance than the homozygous "normals" (CCR5$^+$ CCR5$^+$).

This CCR5 mutation is found at frequencies as high as 8 percent in northern European populations. It shows clinal variation south and east toward Africa and Asia (with frequencies less than 1 percent in those areas). It is virtually absent in African, East Asian, and Native American populations. It is rare in African Americans, its occurrence most likely resulting from admixture during slavery. Given the fact that the HIV virus did not enter Europe to an appreciable extent until the early 1980s, it is impossible for HIV to have been the original selective agent that increased the frequency of this mutation. It is hypothesized that perhaps the individuals who originally migrated to Europe were lucky enough to have high frequencies of this gene (a case of the founder effect) or that some other viral infection faced by European populations may have selected for this particular mutant. It is clear that the distribution of this gene alone is not sufficient to explain the differences in frequency of HIV infection and AIDS in different geographic regions, although it may be playing a role. There are other variables that must be considered. A recent review of the biological and sociological data concerning the African AIDS epidemic concludes that the single variable that best correlates with the spread of AIDS in Africa is the lack of male circumcision.[7] This lack of circumcision, accompanied by widespread poverty, is in turn associated with high rates of chancroid sores. This analysis stands in opposition to some explanations that purport genetically based differentials in sexual behavior and aggression by African males as the main source of the AIDS epidemic in Africa (for example, the explanations of J. Philippe Rushton).[8]

Age-Related Population Genetic Mechanisms

The small overall difference in average allele distribution among human populations does not in and of itself suggest that there might not be genetically determined differences among these populations. Hypothetically, such differences could result from mutations in a pleiotropic gene, that is, a gene that, like a master switch, affects numerous, apparently unrelated characteristics of the phenotype. If different pleiotropic loci were fixed in populations, that could account for widely different rates in numerous sources of mortality. The problem for those who might suggest such a mechanism for explaining racial differences in disease incidence is demonstrating the existence of genes exhibiting pleiotrophy.

The evolutionary theory of aging has demonstrated that genes are typically subject to age-specific expression. This explains why the incidence of biological sources of mortality in human populations generally shows an exponential age-related increase. Until the mid–twentieth century, the explanation for the age-specific increase in biological sources of mortality was obscure. However, in the 1950s Peter Medawar and George C. Williams described the conditions under which alleles that have negative impact on survival late in life accumulate in populations. This theory now rests on a substantial body of mathematical analysis and has been corroborated experimentally in model systems such as the fruit fly *Drosophila*. The relevant biomedical research results were widely disseminated.

Unfortunately, the theory has been widely ignored or misunderstood in most biomedical research related to aging. Molecular reductionism played an important role in the neglect of evolutionary genetic theory in aging and biomedical research. This is unfortunate, as evolutionary theory is the only intellectual program that has successfully explained why metazoans (multicellular animals and plants) age. This program has shown that an allele that has a positive effect on early-life fitness can increase in a population despite its deleterious effect late in life. This has been tested extensively in experimental systems, such as *Drosophila melanogaster,* that have both the predicted phenotypic trade-off between components of early- versus late-life fitness and corresponding changes in gene frequencies that must result from this mechanism. Reviews of the literature on life history in mammals and insects have shown the existence of analogous evolutionary trade-offs within these taxa. We now know that alleles that foster cancer can have positive effects in early life arising from their role in regulating normal growth and differentiation of cells in development. Their promotion of early-life fitness means that such alleles will be fixed (or found at a very high frequency) in all populations. Therefore, we would not expect racially based differentials at loci such as these within our own species. *Since such loci will exist in all populations, it is unlikely that the large differential in cancer incidence and mortality rates in socially constructed races results mainly from such genetic differences.* Given what we know about the distribution pattern of genetic variation as a whole, there is no a priori reason to suspect racially consistent variation for loci such as these; indeed quite the opposite is true.

This is not to say that geographically based polymorphisms that contribute to disease incidence or survivability do not exist. What it does indicate is that these genetic polymorphisms will be distributed in the same way that the rest of the genetic variation in the human species is (that is, independent of the inferred geographic variation that was used to construct racial categories). It is simply not possible to create an unambiguous classification scheme that relates the race of individuals to their probability of disease incidence or mortality. Thus we would expect each subpopulation, owing to its geographic history, to contain unique frequencies of alleles related to disease incidence. These frequencies are determined by the interactions between selection, genetic drift, and mutation.

What Are the Consequences of Associating Race with Disease?

One of the greatest enemies of research is the preconceived biases that scientists unconsciously bring to their work. These biases can come from discipline-specific ways of thinking or from social pressures or from both. In the 1930s Euro-American physicians felt that African Americans were a "notoriously syphilis soaked race."[9] The Tuskegee syphilis study attempted to test the existence of genetic differences in syphilis progression in African Americans as compared to Euro-Americans. More than five hundred subjects were involved in this experiment, most dying before its details came to the public eye. Nothing of scientific value came out of this study. This case reveals both the disciplinary failure of genetic determinism and the white social bias that says that blacks deserved their lot. Genetic determinism led to the assumption that the phenotypic variation examined was the result of an underlying genetic cause, even when obvious and apparent environmental explanations were available. We know that the importance of the interaction between genotype and environment was already well established in evolutionary and quantitative genetics at the time of the syphilis study. The social bias was revealed in the way that whites refused to see their role in creating the economic and social conditions that fostered the extraordinary rates of infection in Macon County, Alabama.

The history of genetics has shown that with each advance in theory and technique there is a rush to apply that advance to explain human variation. Often this application is done recklessly, and that recklessness is directly related to the longevity of the adaptationist assumption. The new molecular reductionism also suffers from this historical trend. The future of the genetic analysis of disease and human variation can be rescued only by the integration of sound evolutionary and population genetic thinking into the paradigm. One aspect of this integration will be the focus on individual and population variation, as opposed to nineteenth-century descriptions of socially defined racial categories.

Cancer incidence shows clear correlation with socioeconomic data, and this correlation is much larger in magnitude than reputed differentials between races. It is extremely difficult to estimate the true genetic components to the variance in incidence rates. The same problem is encountered in all situations where environmental conditions are not controlled. In these circumstances, it is nearly impossible to accurately estimate gene $\times$ environment interaction and covariance of gene and environment effects. Thus, simply identifying gene frequency differences among populations is not a sufficient means of assigning genetic causation to a locus suspected to be involved in disease. This problem has been implicitly recognized in a recent publication of the Institute of Medicine titled *The Unequal Burden of Cancer.*[10]

The problem of environmental variance is not entirely solved by the use of techniques such as the resemblance between relatives or quantitative trait locus mapping. These techniques require large samples of relatives and detailed pedigrees. These are still not entirely free from gene × environment complications. For example, it has been shown that when patients with prostate cancer were uniformly diagnosed and treated similarly, race had no significance as a prognostic variable in the survival of prostate cancer patients. Better medicine, like better science, tends to disprove race-based findings.

What Can or Will We Do without Race?

In the preceding pages I have reviewed the history of the race concept in the Western world. Race as most people understand it now was socially constructed, arising from the colonization of the New World and the importation of slaves, mainly from western Africa. The ramifications of this concept have persisted to this day. The twentieth century was also the period of the greatest increase in scientific knowledge the world has ever known. As the world struggled with issues resulting from its socially constructed views of race, the biological sciences grew at an unprecedented rate. Developments in biochemistry, molecular biology, and population and quantitative genetics created the preconditions for a rigorous examination of human biological diversity. The early studies, initiated in the 1940s, began to seriously discredit previously existing racial taxonomies. During the later decades of the century, data would accumulate opposing the existence of biological races until finally, in the early 1990s, the biological race concept was thoroughly dismantled.

Yet, despite these scientific advancements, we begin this new millennium with most Americans unaware of the implications of these new scientific facts. For example, an ABC News poll released on October 4, 1999, suggested that Americans currently see racism, prejudice, and the hate crimes they engender as some of the most pressing problems to be addressed in the new millennium.[1] These responses may have been influenced by the rash of racist hate crimes that occurred in 1999 (the Columbine and Granada Hills shootings both targeted racial minorities). The confused idea that observable biological differences such as skin tone and hair type are accurate predictors of personality, intellect, and morality is a core component of racist belief systems. Thus, a crucial part of the battle against the legacies of the social construction of race is to get across the message that biological races do not exist and that these types of correlations are spurious.

We are at a crossroads concerning racial attitudes in the United States. To return to our central metaphor, the emperor's parade is progressing down the avenue and everyone suspects that something is amiss, but we still have not heard or understood the child's critical message: the emperor race is naked. It is crucial for the scientific community to actively translate this message to the American public. We need to aggressively present the broad consensus proclaiming the lack of validity of the biological race concept in every appropriate venue. Some may argue that the scientific community cannot accomplish or should not be appointed to accomplish this task. After all, to what degree can scientists influence public opinion on crucial issues? Finally, they might ask, How would American society change if everyone understood these points?

In the first regard, the ultimate power of science results from its ability to make sense of the natural universe. Surely when Copernicus put forward his theory of the heliocentric solar system, most Europeans believed that the universe was Earth centered. Indeed, it was not just the general population that believed this; powerful social institutions like the church did also. Opposition to this institution was punishable by death; Giordano Bruno was executed in part for his heliocentrism, and Galileo was tortured and imprisoned. However, eventually the fact that the heliocentric solar system model better explained the movement of the planets led to the abandonment of the geocentric model. The adoption of this more correct theory of the solar system paved the way for the accumulation of even greater knowledge, with important utilitarian effects. The biological race concept, which is demonstrably false, must be replaced with our more accurate understanding of biological diversity within our species. Once this is accomplished we can move forward and create new and useful knowledge. Scientists have shown that they can influence public thinking. For example, Rachel Carson's *Silent Spring* (1962) created a new awareness of the environment in the American public. Few would argue that this book did not profoundly influence the way Americans viewed their environment and that this did not spur people to mass collective action to better protect it. I hope that my work will stimulate this kind of reassessment of the biological race concept. The United States' ongoing confusion about race results not from any intrinsic difficulty explaining the topic of human genetic variation but from the deep entanglement of racial categorization in the historical and social fabric of the country. Success in clarifying this issue will result only from intellectual courage, resolve, and resourcefulness.

An important step toward this goal will be to take this message out of the universities. Educators have not sufficiently infused the K–12 curriculum with the modern scientific understanding of human biological variation. In 1999, several states (Kansas, Kentucky, and New Mexico) actually weakened our ability to accomplish this end by removing any explicit reference to evolution from their state science standards. Evolutionary biology must be a core requirement if students are to learn the correct explanation for human genetic variability. Removing evolution from state science standards means that K–12 teachers will not necessarily be expected to be proficient, and those who are may fear to address controversial applications of the theory in the classroom.

In addition, most students who enroll in university courses are never exposed to discussions of the significance of human genetic diversity. Undergraduate students rarely take courses in anthropology or evolutionary biology. Most universities do not require that students enroll in courses that focus on cultural diversity. Finally, ethnic or African American studies programs are generally not prepared to adequately present the science behind the race concept. This condition persists simply because most universities lack faculty members trained in the required disciplines. Fewer still have a core set of courses capable of taking on the entire scientific and philosophical program behind biological racism. I hope that I have made a case for such courses to be developed at all appropriate levels of instruction.

The fallacy of the biological race concept must be incorporated into our collective thinking on an everyday basis. For example, dictionaries and encyclopedias need to be revised to include comprehensible and correct definitions of "race" and explanations of human genetic diversity. We also must begin to talk about our own identities outside of the racial paradigm. We must build a new common language that accurately describes individuals within our populations. We must abandon the practice of describing ourselves as "black," "brown," "red," "white," or "yellow." In particular, historically oppressed populations should stop describing themselves utilizing such worn euphemisms such as "minorities" or "persons of color." Furthermore, federal and state agencies should stop using these incorrect and misleading terms. For example, the federal Office of Management and Budget (OMB) categories describing individuals are hopelessly confused. One of the most misleading practices of the OMB is the continued use of the term "white" as a racial category. The other four categories have some geographic designation in their titles, American Indian/Alaska Native, Asian, African American, and Native Hawaiian/Pacific Islander. Why are Asians not Asian Americans and whites not European Americans?

There is a simple and rational way to define individuals whose immediate ancestry defies the socially defined categories. Simply permit them to check the blocks on the census that allow them to describe their ancestry as accurately as possible. The 2000 census forms allowed double checking, but those who double checked were counted in the "darker" category between "white" and minority.[2] The OMB's decision to count this way resulted from the concern expressed by some civil rights organizations that "minorities" would be undercounted. The OMB justified this decision by claiming that people who had suffered historical discrimination should be subject to certain "protections." However, the decision of the Clinton administration on this issue reinforces the American tradition of hypo-descent and ideologically supports continued segregation. One could argue that one important measure of the United States' movement toward an antiracist society would be the growth of its "multiracial" category. The OMB now recognizes sixty-three different racial combinations. These combinations will better describe the complexities of American social life in the twenty-first century, and the number of individuals who consider themselves of multiracial ancestry should be reported in each census by category. This ongoing confusion concerning biological and social conceptions of human diversity testifies to the continuing reign of the emperor race.

How Will We Benefit by Dismissing Racial Thinking?

It is not enough, however, to simply dismiss the biological category of race. The questions that I have raised indicate that we must determine the practical consequences of this dismissal for the social debate. The United States still suffers from the huge political and economic disparities between those derived from northern European ancestry and those who are not. How might the story have been different if we had recognized that there are no races in the human species? What if we had started out by recognizing the significance of the fact that genetic diversity is greater within racial groups than between them? We know that family lineages within geographic populations exhibit different gene frequencies. These, under the influence of environmental variation, determine the value for both qualitative and quantitative traits associated with stamina, height, personality, and even intelligence (however we define it). In this scenario, we would expect that a society that brought together different geographical populations would not be stratified by race. Instead, we would expect diversity among those who became successful. The presidency, the Congress, the Supreme Court, and the financial and industrial complex would reflect that diversity. We would also expect that the original geographical differences would have been rapidly homogenized. There would have been no reason to maintain marital prejudices in such a society. It would have been recognized that the few genetic loci that determine skin color or hair type were in principle no different from those that determine blood or fingerprint type. In this story, if humans had insisted on the need for prejudice they would have had to find some other traits. Maybe religion, gender, or sexual preference could have been used, but race would have been eliminated.

The problem as we enter the twenty-first century is that we have believed in the reality of race for so long that many will wonder how to live without it. Socially constructed race seems as tangible as gravity. Ironically, a new school of racism has developed that denies the significance of the social construction of race. Utilizing the idea that there is no longer any discrimination in our society, this new school claims that to be concerned with the United States' racial legacy is to be a racist.[3] *Clearly, recognizing that no biological races exist in our species cannot be confused with claiming that socially defined racism has not existed and is not still a problem.* The purpose of this book was to remove the scientific foundations of claims that genetic or biological differences in human populations determine their social conditions. To recognize that there is no legitimate way to classify people into races is to recognize that all the social ills and benefits can be found among all populations. This means that we must find a way to begin to treat people as the unique individuals they are. This does not mean that we can abandon efforts to bring about social equity between populations that have historically been denied equal opportunity. Thus, the fact that there are no biological races in the United States is not a reason to end programs designed to remedy past discrimination. It is precisely this recognition that makes it imperative that we reverse the environmental deprivations that continue to destroy lives and divide us as a nation. The difficult question is, How can we design programs that progressively eliminate the detriments caused by the history of racist injustice and yet simultaneously defend the rights of all individuals?

This is the charge the United States must take up in the new century. A cornerstone in this struggle will be the wide dissemination of the fact of the non-existence of biological races. This fact will not be sufficient to end racism by itself. Knowledge of truth does not guarantee action. Ultimately, the United States will need a moral rebirth to accomplish the final destruction of racist ideology and practice. What we require now is a transitional program that recognizes the legacy of racism and identifies what we will gain by eliminating thinking that is still based on racial categories.

Two Futures

How might we accomplish this in practice? First we need to integrate the new understanding of genetic diversity within our species into our social institutions. No one should assume that the genetic history of a population determines the social, cultural, or personality traits of an individual. As a young child, I was tracked into a curriculum designed for those expected to have less intellectual ability. In third grade, when I began to read books about the history of the Crusades, my teachers felt that I was compensating for my inability to read (the other children were still reading picture books with large print). Finally, after one teacher decided to determine just how much I understood in this book, the teachers realized that they were dealing with a child of exceptional rather than limited ability. The school system had assumed that because I was "colored" and my parents were poor, my learning potential was limited.

My childhood experience is also evidence against the assumption that the environmental history of a population uniquely determines the characteristics of an individual. We should try to become as sensitive as we can in our public and private institutions to individual differences. We cannot assume that all rich children will be smart and all poor children will be intellectually inadequate. We do, however, need to recognize that there are infrastructural requirements for learning, such as adequate instruction, decent nutrition, safety from intimidation and violence, and an environment that fosters the development of self-esteem. For example, research has shown that racist stereotypes impact how and what students believe they can learn.[4] Furthermore, we know that neglect and verbal abuse can actually alter the brains of those who are chronically exposed to such abuse. This abuse leads to lower cognitive function and personality disorders. It is well known that environmental toxins do similar damage. I have established in this book that these negative environmental parameters are disproportionately visited upon minority populations (for example, African Americans, American Indians, and Hispanics). Conversely, positive environments encourage the development of positive cognitive abilities and personality traits, and even physical changes in the brain consistent with these abilities. The experiments of Berkeley neuroanatomist Marian Cleeves Diamond are particularly relevant in this regard. She has been able to show that exposing genetically identical rats to poor environments actually produced different brain structures than exposing them to rich environments. Poor environments were those without toys or objects to stimulate the rats' curiosity, whereas rich environments gave rats access to these objects. Poor-environment rats had smaller neurons with few dendritic connections, whereas rich-environment rats showed larger neurons with greater dendritic connections. A recent examination of Albert Einstein's brain showed that he had these characteristics in the brain region associated with mathematical ability.[5] The new neuroscience suggests not that he was born that way but that his continued interest in mathematical problems created positive feedback for the neural ability to accomplish in the field. It seems then that our real challenge for the twenty-first century is creating environments that allow for the healthy expression of the genetic variability that exists in our nation.

If we accomplish this, then we can begin to address the operational racist dysfunction in our society. For example, by addressing the concerns raised in the introduction, we could begin to use the resources of all individuals (instead of those of just a few). How many potential Ben Carsons (Carson is an African American neurosurgeon at Johns Hopkins University) are lost to the despair caused by racist injustice? In addition, eliminating racist discrimination would further diversify the ethnic and cultural backgrounds of the United States' political leadership. Solutions to social problems such as poverty, delinquency, and crime might benefit from having new perspectives at the table. Groups that have always felt disenfranchised might be more willing to cooperate to solve these problems, particularly if they felt that their voices were really being heard. Changes such as these would also facilitate communication between groups and help to disseminate accurate knowledge of minorities and their culture to society at large. The end of racial discrimination would also promote respect for law enforcement and for the peaceful settlement of disputes. Thus, there are clear benefits to eliminating the effort required to maintain the barriers that prevent the full participation of all members of society. Ask yourself, What might be the impact on African, Hispanic, Asian, or Native Americans if a member of one of these groups were elected president? Finally, if racial prejudice and discrimination do undercut goodwill and friendly diplomatic relations between nations, their elimination might increase global trade. Economic growth is likely to be beneficial to all parties.

On Tuesday, August 10, 1999, Buford O. Furrow Jr., a self-confessed white supremacist, walked into a day care facility at a Jewish community center and opened fire on defenseless children and employees. After fleeing from the scene, Furrow found and murdered a Filipino man, postal carrier Joseph Ileto, because he was not white and because he worked for the federal government. Furrow had ties to hate groups and a history of mental illness. The police found a small arsenal of weapons and ammunition in his vehicles. The rash of hate crimes in 1999 resulted from the idea held among white supremacists that the millennium would signal the start of the final race war. These cowardly acts of terror are designed in their minds to be a wake-up call to white Christians to defend themselves from the international Jewish conspiracy and its "colored" foot soldiers.

We are now faced with two futures, one envisioned by the Buford Furrows of the world, the other articulated by the late Reverend Martin

Luther King Jr. Will all God's children's ever join hands? Whether they do so or whether we continue to allow the fallacies embodied in the biological race concept to dictate our futures is entirely up to us. *The Emperor's New Clothes* was also written as plea for us to open our eyes. We have lived in the nightmare of racism too long. It is time to put an end to it. No one should have to fear for his or her life because of ethnicity or skin color. The real problem, however, is not the visible hate crimes of extremists like Timothy McVeigh, Benjamin Nathaniel Smith, and Buford Furrow but the systematic injustice that breeds their ideological justification. There can be no race war if there are no races. The conflict then is really between socially constructed cultural groups. The conflict resides not in some immutable difference rigorously coded by our genes but in our social institutions. We can change these institutions; we need only have the political and moral will to do so. To begin, we must listen to that child who sees so clearly: "Look, the emperor race is wearing no clothes."

Admixture of Genes in Human Populations

To illustrate how genetic admixture may make human populations more similar than they outwardly seem, we can examine the genes at the tissue compatibility loci, called the histocompatibility loci antigen genes (HLA-A, HLA-B, and HLA-C). The HLA loci are responsible for the body's ability to identify foreign proteins and carbohydrates, an ability that is crucial to soft-skinned organisms, which are subject to routine invasion by microbes. To resist invasion by microbes, the HLA genotypes must be complex (that is, there must be many possible genotypes). The necessity for complexity explains why the HLA loci have many alleles compared to a locus such as that for eye color, where only a few options reside. Alleles ultimately refer to portions of DNA that code for a protein, and alleles may be identified in many ways (for example, by protein, or now DNA, electrophoresis).

There are at least forty-four alleles reported at the HLA-B locus.[1] The frequency of the bw61 and bw72 allele variants at this locus in Africans and in three major North American racial groups is shown in table A.1. This table shows that on average we would expect 110/1000, 26/1000, 125/1000 whites, blacks, and American Indians, respectively, to have the bw61 allele. The absence of bw61 or bw72 in Africa suggests that its presence in North American blacks must have resulted from admixture with either whites or American Indians.

Table A.2 examines the frequencies of the cw8 allele at the HLA-C histocompatibility locus. Again, these frequencies suggest that cw8 must have arisen from whites, since it is absent in Africans and American Indians.

We must exercise caution in interpreting these data. First, we must recognize that the racial categories used in this example were socially constructed. The calculation of gene frequencies at this locus might differ if we chose different criteria to construct our groups. We might, for example, see differences in gene frequencies of the same magnitude if we divided our sample into a group of individuals with loops in their fingerprints and a group with whorls. The frequencies in these new categories might result because the variation in HLA frequency is not determined simply by geographic origin. We might also ask how the individuals were identified. A person of nine-sixteenths European and seven-sixteenths African ancestry might appear black, was probably raised black, and probably identifies himself or herself as black. How would

Table A.1. Allele Frequencies at the HLA-B Histocompatability Locus

	bw61	bw72
Africa	0	0
North America		
Whites	0.11	0
Blacks	0.026	0.079
American Indians	0.125	0.021

SOURCE: A. K. Roychoudhury and M. Nei, *Human Polymorphic Genes: World Distribution* (New York: Oxford University Press, 1988), 150–153.

Table A.2. Allele Frequencies at the HLA-C Histocompatability Locus

	cw8
Africa	0
North America	
Whites	0.055
Blacks	0.024
American Indians	0

SOURCE: A. K. Roychoudhury and M. Nei, *Human Polymorphic Genes: World Distribution* (New York: Oxford University Press, 1988), 237.

researchers classify this person? There has also been a historical neglect of non-European populations in research related to health care.

The estimated frequencies of these alleles result from studies designed to examine tissue compatibility for organ donation in various populations. There have been fewer studies from western Africa, with fewer individuals sampled, and from American Indian populations. Therefore, estimations of gene frequencies in these populations are likely to be less accurate.

When we estimate the amount of unidirectional gene flow at the HLA loci into North American blacks, we find that the migration rate (m) of this allele into the African American population is equal to 5.6 percent per generation. Another case of admixture can be seen from the frequency of alleles at the MN blood group locus. For example, allele frequencies were measured for African Americans in Claxton, Georgia, and the frequency of allele M was found to be 0.484. For Euro-Americans in Claxton the M frequency is 0.507, whereas in West Africa the frequency of the allele M is 0.474. On the basis of these figures, geneticists assumed that the modern Claxton African Americans are descendants of West Africans. Therefore their gene frequencies should be similar to those observed in those populations. There are many simplifying components to this assumption, but they are not important enough to raise here. Again, we can calculate the rate of migration (m) of European genes into the African population; for this locus the rate is 0.035 per generation.[2]

Table A.3. Estimates of Unidirectional Gene Flow from Euro-Americans to African Americans

Allele	Rate per generation
Ss blood groups	-0.013
Duffy blood groups	0.011
Kidd blood groups	-0.028
Kell blood groups	-0.005
G6PD deficiency	0.039
Hemoglobin-B	0.071

SOURCE: Data are from D. Hartl and A. G. Clark, *Principles of Population Genetics*, 2d ed. (Sunderland, Mass.: Sinauer Associates, 1989), 308.
Note: A negative sign could mean that the allele increased in whites, suggesting that gene flow was not unidirectional.

The migration rate has been estimated for other loci (see table A.3). Additional studies of the Duffy blood group (Fy^a) allele have estimated the rate of admixture from Europeans to Africans in the United States. Like the cw8 allele, this allele is virtually non-existent in West Africa. The estimate using that locus was about 1 percent admixture per generation. However, other estimates of the admixture problem give 3.5 percent per generation (utilizing blood group proteins, such as RH, FY, or GM globulins). Thus it is generally estimated that as much as 30 percent of the African American gene pool is derived from Caucasoid or Mongoloid origins.

The Calculation and Significance of Genetic Distance in Human Populations

The analysis of genetic variability in human populations began as part of the general question concerning how much genetic variability we should expect in nature. In 1966, R. C. Lewontin and J. L. Hubby utilized protein electrophoresis to examine the amount of existing genetic variability in *Drosophila pseudoobscura* populations. Classical population genetic theory predicted that there should not be much standing genetic variability in populations; the theory suggested that at most loci one wild-type allele should be favored, mutants should occur at very low frequency, and heterozygosity should be low. Electrophoretic observations showed that heterozygosity was much higher than can be accounted for by the classical theory. A similar phenomenon is observed in humans. For example, 33 percent of all human loci surveyed were found to be polymorphic (a polymorphic locus is defined in population genetics as having mutant alleles at a frequency greater than 1 percent). By extension, this means that 67 percent of human loci are generally invariant (except for the presence of rare mutations) and therefore cannot contribute to major genetic differences between populations.

One of the most interesting results of the early studies was that the amount of genetic variability at polymorphic loci within geographically separated populations of *Drosophila* was greater than the variability between populations. Thus, utilizing the standard statistical criteria for the evaluation of population difference, such populations could not be distinguished as "different" genetically. This same phenomenon was also immediately demonstrated in human populations. Masatoshi Nei and his co-workers confirmed the fruit fly results by examining the polymorphic genetic variation within and between the so-called three major races of man over a period of sixteen years. The genetic variation between the so-called human races has been reported using a variety of methods, including analysis of general proteins, blood groups, histocompatibility locus antigens (HLA), and DNA markers (see table B.1). In addition, two studies analyzing the similarity of restriction site differences for mitochondrial DNA have been reported (the mitochondria are the powerhouse organelles within the cell; see table B.2). Mitochondrial DNA is inherited only through maternal lineages. Restriction enzymes are found in bacteria, and they recognize specific sites within the DNA. They cleave the DNA at these

Table B.1. Mean Genetic Distances and Standard Errors among the Three Major Ethnic Groups, Based on 186 Polymorphic Loci

Genetic loci	No. of loci	Europeans-Asians	Europeans-Africans	Asians-Africans
Proteins	84	0.028±0.009	0.035±0.009	0.048±0.012
Blood groups	33	0.019±0.010	0.059±0.032	0.082±0.041
HLA and immunoglobulins	8	0.329±0.122	0.701±0.341	0.386±0.169
DNA markers	61	0.060±0.012	0.061±0.017	0.109±0.025
Total	186	0.040±0.007	0.063±0.011	0.078±0.013

SOURCE: M. Nei and G. Livshits, "The Genetic Relationships of Europeans, Asians, and Africans and the Origin of Modern *Homo sapiens*" *Human Heredity* 39 (1989): 276–281.

Table B.2. Genetic Distances and Standard Errors Based on Mitochondrial DNA Restriction Differences

	Europeans-Asians	Europeans-Africans	Asians-Africans
Cann 1987[a]	0.02	0.05	0.04
Brown 1980	0.032±0.046	0.045±0.069	0.036±0.059

SOURCE: R. Cann, M. Stoneking, and A. C. Wilson, "Mitochondrial DNA and Human Evolution," *Nature* (Lond.) 325 (1987): 31–36; and W. M. Brown, "Polymorphism in Mitochondial DNA of Humans as Revealed by Restriction Endonuclease Analysis," *Proceedings of the National Academy of Sciences USA* 77, no. 6 (1980): 3605–3609.
[a]Standard errors unavilable for Cann data.

sites, and treatment with a standard group of restriction enzymes will produce a characteristic pattern for a specific DNA sequence.

A simple way to interpret these results is to ask, What are the chances that two unrelated individuals from purportedly different races share the same allele at a polymorphic locus? This chance is approximately equal to 1.0 minus the average genetic distance between the populations. Thus Africans and Europeans would have about a 93.7 percent chance by total protein analysis or as much as 96.4 percent chance by mitochondrial DNA analysis of sharing a given allele. We know that there are about 80,000 proteins that are encoded in the human genome. Of these, approximately 23,000 are polymorphic, and of these about 1,000 would differ on average between Europeans and Africans. Among these 1,000 loci, there will be some genes of major impact and many of minor quantitative impacts. It is possible that some important component of complex behavioral traits could be coded for by a small number of loci.

Genetic distances between populations may be calculated using measures of genetic similarity or difference between populations X and Y. For example, Sewall Wright's 1951 measure of genetic distance was called F_{ST}:

$$F_{ST} = V_p/P_{avg} (1 - P_{avg}),$$

where V_p = the variance between gene frequencies of a set of n populations and P_{avg} = their average gene frequency. Other commonly used measures are Nei's index of genetic similarity (I_N), Nei's index of genetic difference (D_N), and Rogers's index of genetic similarity (S_R):

$$I_N \text{ (for locus, i = 1, m)} = \Sigma(p_{ix} * p_{iy}) / [(\Sigma p_{ix}^2) (\Sigma p_{iy}^2)]^{1/2}$$

$$D_N = -\ln(I_N)$$

$$S_R = 1 - [1/2 \Sigma(p_{ix} - p_{iy})]^{1/2},$$

where p_{ix} = frequency of allele i in population (or species) X and p_{iy} = frequency of allele i in population (or species) Y.

All these measures utilize some function of the average and variance of given alleles found within the populations in question. Best results for determination of genetic distance are obtained when a large number of loci are examined and alternative measures of genetic distance are used. The results shown in table B.1 are robust when these conditions are met.

Notes

Preface

1. The papers were published in "Genetics for the Human Race," *Nature Genetics Supplement* 36, no. 11 (November 2004).

2. Nicolas Wade, "Articles Highlight Different Views on the Genetic Basis of Race," *The New York Times* (October 27, 2004); Armand Leroi, "A Family Tree in Every Gene," Op-Ed contribution, *The New York Times* (March 14, 2005).

3. I also discuss this in the preface to the paperback edition of my book, *The Race Myth: Why We Pretend That Race Exists in America* (New York: Dutton Books, 2005).

4. M. Marchione, "Heart Failure Drug for Blacks Expected to Become First Pill Sold for a Specific Race," Associated Press (November 9, 2004).

5. A. Taylor et al., "Combination of Isosorbide Dinitrate and Hydralazine in Blacks with Heart Failure," *The New England Journal of Medicine* 351, no. 20 (2004): 2049–2057.

6. E. S. Epel et al., "Accelerated Telomere Shortening in Response to Life Stress," *Proceedings of the National Academy of Sciences* 101, no. 49 (2004): 17312–17315.

7. W. Chen et al., "Nitric Oxide Synthase Gene Polymorphism (G894T) Influences Arterial Stiffness in Adults: The Bogalusa Heart Study," *American Journal of Hypertension* 17, no. 7 (2004): 553–559.

8. E. S. Epel et al. (2004).

Introduction. Racial Thinking

1. Indications that many Americans still believe that innate racial differences exist can be taken from A. Almquist and J. E. Cronin, "Fact, Fancy, and Myth in Human Evolution," *Current Anthropology* 29, no. 3 (1988): 520–522; and also, more recently, "Colorblind: Talking about Race under Internet Cover," MSNBC series, April 1998.

2. L. L. Cavalli-Sforza, P. Menozzi, and A. Piazza, *The History and Geography of Human Genes* (Princeton, N.J.: Princeton University Press, 1994).

3. R. Holmes, *How Young Children Perceive Race*, Sage Series on Ethnic and Race Relations, vol. 12 (Thousand Oaks, Calif.: Sage Press, 1995).

4. See the arguments for dysgenesis presented by R. J. Herrnstein and C. Murray, *The Bell Curve: Intelligence and Class Structure in American Life* (New York: The Free Press, 1994), 343–357; and R. Lynn, *Dysgenics: Genetic Deterioration in Modern Populations* (Westport, Conn.: Praeger, 1994).

5. P. Gagneux, C. Wills, and U. Gerloff, "Mitochondrial Sequences Show Diverse Evolutionary Sequences of African Hominids," *Proceedings of the National Academy of Sciences USA* 96 (1999): 5077–5082.

6. After R. Schaeffer, *Racial and Ethnic Groups,* 7th ed. (New York: Longman, 1998), 14–15.

Chapter 1. The Earliest Theories

1. Modern evolutionary genetic analysis suggests that the human species began in Africa, and thus the idea that all modern humans are descended from Noah's descendants is not supported. However, there may be at least one group of South Africans, the Lemba tribe, that can trace its ancestry to Israel (much later than the story of Noah, however). The evidence supports the idea that this tribe has a male lineage that can be traced to Israel; see, for example, M. G. Thomas, T. Parfitt, D. A. Weiss, K. Skorecki, J. F. Wilson, M. le Roux, N. Bradman, and D. B. Goldstein, "Y Chromosomes Traveling South: The Cohen Modal Haplotype and the Origins of the Lemba, the 'Black Jews of Southern Africa,'" *American Journal of Human Genetics* 66 (2000): 674–686. This story also underscores the fallacy of race, in that the Lemba claims of Jewish ancestry were dismissed because the Lemba were sub-Saharan Africans. This can be explained, however, by the marriage of male migrants from the Middle East, most likely Yemen, into the native population of sub-Saharan African.

2. References to the oral tradition in the Talmud are found in *The Jewish Encyclopedia* (New York: Ktav Publishing House, 1925), 6:186.

3. F. Snowden, *Blacks in Antiquity: Ethiopians in the Greco-Roman Experience* (Cambridge: Belknap Press, 1970), 201–205.

4. The kingdom of Sheba and its queen have often been claimed as part of the Arabic world. It has been suggested that she was instead a queen of a kingdom in Africa, and the queen herself is described as Ethiopian. The writings of Flavius Josephus, *Antiquitates Judaicae,* refer to her as a queen of Egypt and Ethiopia. Edward Ullendorff notes that the Josephus passage apparently referred to Nubia-Meroe rather than to Abyssinia, modern day Ethiopia. Josephus (A.D. 37 or 38–ca. 101) was a Jewish historian, born in Jerusalem of both royal and priestly lineage. His original name was Joseph ben Matthias. See Snowden, *Blacks in Antiquity,* 201–205; and E. Ullendorff, "The Queen of Sheba," *Bulletin of the John Rylands Library* 45 (1962–1963): 492–493.

5. T. Gossett, *Race: The History of an Idea in America* (New York: Schocken Books, 1963), 8–9.

6. Thomas Gossett (ibid., 7) states that there is no consistent relationship between membership in higher castes and physical features, whereas J. Philippe Rushton claims that the Indian caste system was highly successful in maintaining physical types, until recently (J. P. Rushton, *Race, Evolution, and Behavior:*

A Life History Perspective [New Brunswick, N.J.; Transaction Publishers, 1995], 92). There is now some evidence that the Indian caste system did create genetic differences where none existed before; see, for example, D. Govindaraju, "Our Tangled Heritage: Demographic History and the Indian Caste System," *American Journal of Human Genetics* 57 (1995): 5.

Chapter 2. Colonialism, Slavery, and Race in the New World

1. C. Darwin, *The Descent of Man, and Selection in Relation to Sex* (London: J. Murray, 1871; reprint, Princeton, N.J.: Princeton University Press, 1981), 199: "We are naturally led to inquire where was the birthplace of man at that stage of descent when our progenitors diverged from the Catarhine stock. . . . In each great region of the world the living mammals are closely related to the extinct species of the same region. It is therefore probable that Africa was formerly inhabited by extinct apes closely allied to the gorilla and chimpanzee; and as these two species are now man's nearest allies, it is somewhat more probable that our early progenitors lived on the African continent than elsewhere." Darwin's intuition concerning the closeness of Catarhine apes to humans was later verified by Morris Goodman; see his "Immunochemistry of the Primates and Primate Evolution," *Annals of the New York Academy of Sciences* 102 (1962): 219–234. That this evolutionary closeness to Old World apes indicates Africa as the original home of modern humans is verified by the work of the Leakeys and Johansons; see, for example, R. Leakey and R. Lewin, *Origins Reconsidered: In Search of What Makes Us Human* (New York: Anchor Doubleday, 1992); and D. Johanson, L. Johanson, and B. Edgar, *Ancestors: In Search of Human Origins* (New York: Villard Books, 1994).

2. P. Fryer, *Staying Power: The History of Black People in Britain* (London: Pluto Press, 1984), 139–140. Also see E. Jones, "The Physical Representation of African Characters on the English Stage during the Sixteenth and Seventeenth Centuries," *Theatre Notebook* 17 (1962–1963): 18–19. Jones points out that there are at least thirty-six plays and masques in which African characters are portrayed mostly negatively in this period.

3. Francis Bacon, "New Atlantis: A Worke Unfinished, 26," in *Sylva sylvarum, or, A Natural History in Ten Centuries* (London: W. Lee, 1627).

4. H. Zinn, *A People's History of the United States* (New York: Harper Colophon Books, 1980), 15.

5. W. Rodney, *How Europe Underdeveloped Africa* (Washington: Howard University Press, 1974); and S. Amin, *Neo-Colonialism in West Africa,* trans. Francis McDonagh (New York: Monthly Review Press, 1973). For a discussion of the impact of the slave trade on fostering European colonization of Africa, see P. Manning, *Slavery and African Life: Occidental, Oriental, and African Slave Trades* (Cambridge: Cambridge University Press, 1990), 30–31.

6. The data are from http://www.whc.neu.edu/simulation/histexmp.html, a Web site maintained by Professor Patrick Manning of the Department of History and African-American Studies at Northeastern University; additional references supporting these data come from Manning, *Slavery and African Life;*

Manning, "Slave Trade: The Formal Demography of a Global System," *Social Science History* 14 (1990): 255–279; M. Patrick and W. Griffiths, "Divining the Unprovable: Simulating the Demography of African Slavery," *Journal of Interdisciplinary History* 19 (1988): 177–201; and Manning, "The Enslavement of Africans: A Demographic Model," *Canadian Journal of African Studies* 15 (1981): 499–526.

7. L. L. Cavalli-Sforza, P. Menozzi, and A. Piazza, *The History and Geography of Human Genes* (Princeton, N.J.: Princeton University Press, 1994), 79–80.

8. R. Peterson, *Cages to Jump Shots: Pro Basketball's Early Years* (New York: Oxford University Press, 1990), 119.

Chapter 3. Pre-Darwinian Theories of Biology and Race

1. "Negro Tom," described as a self-taught mathematical genius, was born in Africa and learned without instruction the ability to do complicated arithmetical calculations. The original report is given in B. Rush, "Account of a Wonderful Talent for Arithmetical Calculation, in an African Slave, Living in Virginia," *American Museum* 5 (1789): 62–63.

2. Jefferson's views on Banneker are described in Lerone Bennett Jr., *Before the Mayflower: A History of the Negro in America* (New York: Penguin Books, 1964), 66; and his disregard for the intellectual abilities of Phillis Wheatley is discussed by Thomas Gossett in his *Race: The History of an Idea in America* (New York: Schocken Books, 1963), 43.

3. There is now DNA evidence suggesting that Jefferson fathered at least one of Sally Hemings's children; E. Foster, M. A. Joblina, and P. G. Taylor, "Jefferson Fathered Slave's Last Child," *Nature* 396, no. 6706 (1998): 27–28.

4. S. J. Gould, *The Mismeasure of Man* (New York: Norton, 1981), 50–69.

5. There is abundant historical documentation of scalp hunting and bounties for American Indian populations; see, for example, D. Wrone and R. Nelson, eds., *Who's the Savage?* (Malabar, Fla.: Krieger Publishing, 1982), 96–98; and D. Owsley and R. Janz, *Skeletal Biology of the Great Plains* (Washington, D.C.: Smithsonian Press, 1994), 337. In addition, differential mortality of American Indian populations due to warfare, disease, and relocation could have accounted for the greater availability of Indian skulls; see, for example, H. Zinn, *A People's History of the United States* (New York: Harper Colophon Books, 1980), 124–146.

6. S. J. Gould, *The Mismeasure of Man*, revised and expanded ed. (New York: W. W. Norton, 1996), 96, table 2.4.

7. Rushton's discussion of table 2.4 is found in J. P. Rushton, *Race, Evolution, and Behavior: A Life History Perspective* (New Brunswick, N.J.: Transaction Publishers, 1995), 114–116.

8. Bernard Ortiz de Montellano has shown that Egypt was neither a "Caucasian" society, as claimed by Morton and other nineteenth-century racists, nor solely a sub-Saharan African society, as claimed by many modern Afrocentrists; see Ortiz de Montellano, "Melanin, Afrocentricity, and Pseudoscience," *Year-*

book of *Physical Anthropology* 36 (1993): 33–58; and Ortiz de Montellano, "Multiculturalism, Cult Archaeology, and Pseudoscience," in *Cult Archaeology and Creationism: Understanding Pseudoscientific Beliefs about the Past*, ed. F. B. Harrold and R. A. Eve (Iowa City: University of Iowa Press, 1995).

9. F. Douglass, "The Claims of the Negro Ethnologically Considered," in *The Life and Writings of Frederick Douglass*, Vol. 2, *Pre–Civil War Decade*, ed. P. Foner (New York: International Publishers, 1952), 298.

Chapter 4. Darwinism Revolutionizes Anthropology

1. R. Basler, *Abraham Lincoln: His Speeches and Writings* (New York: World Publishing, 1946), 301.

2. Ibid., 303.

3. A. Desmond and J. Moore, *Darwin* (London: Michael Joseph, 1991), 155.

4. James Hunt, "Address Delivered at the Third Anniversary Meeting of the Anthropological Society of London," *Journal of the Anthropological Society of London* 4 (1866): lxxviii.

5. Desmond and Moore, *Darwin*, 442–443.

6. F. Burkhardt and S. Smith, eds., *Correspondence of Charles Darwin* (Cambridge: Cambridge University Press, 1985), 6:184.

7. C. Darwin, *The Descent of Man, and Selection in Relation to Sex* (London: J. Murray, 1871; reprint, Princeton, N.J.: Princeton University Press, 1981), 215.

8. Ibid., 231.

9. Ibid., 232.

10. Ibid., 249.

11. Ibid., 217–218, footnote.

Chapter 5. Race and Social Darwinism

1. A. Desmond and J. Moore, *Darwin* (London: Michael Joseph, 1991), 521.

2. Ashley Montagu, *Man's Most Dangerous Myth: The Fallacy of Race*, 5th ed. (New York: Oxford University Press, 1974), discussed the basis under which the racial classification of Negro varied by state. The general rule (as described earlier) was that any individual with any detectable African ancestry was considered a Negro. W.E.B. Du Bois, *Black Reconstruction* (New York: Atheneum, 1935), suggested that by 1860, the African American was on average about 20–30 percent European and 10 percent American Indian by admixture. These figures were later verified by population genetic studies as reported in L. L. Cavalli-Sforza, P. Menozzi, and A. Piazza, *The History and Geography of Human Genes* (Princeton, N.J.: Princeton University Press, 1994). Despite this, Arkansas defined anyone with a "drop" of Negro blood in them as Negro; in Florida, if you had less than one-eighth Negro blood, you ceased to be a Negro; and in Oklahoma, anyone who was not of Negro blood was classified as white.

Even today, under Louisiana law, anyone less than one-thirty-second African descent is considered white. Consider, the famous recent case in which a reputedly white woman had to sue to have her racial status reversed (she was found to be one-thirty-second African American and classified as black); see M. Omi and H. Winant, *Racial Formation in the United States: From 1960 to the 1990s* (New York: Routledge, 1994). Also, in 1963 Mississippi banned white-Mongolian marriages, but North Carolina allowed them. Several landmark Supreme Court cases on white-black marriages were fought even into the 1970s.

3. The *Atlantic Monthly* article praising Spencer is discussed in R. Hofstader, *Social Darwinism in American Thought* (Boston: Beacon, 1992), 33.

4. Ibid., 31.

5. The army study of 130 infected blacks, 632 infected whites, 2,877 uninfected blacks, and 5,625 uninfected whites is reported in R. M. Yerkes, "Psychological Examining in the United States Army (1921)," cited in A. Chase, *The Legacy of Malthus: The Social Costs of the New Scientific Racism* (New York: Knopf, 1977), 638. These tests are also discussed in different contexts in chapters 7, 9, and 10.

6. C. Darwin, *The Voyage of the* Beagle (1845; London: J. M. Dent & Sons, 1906), 481.

Chapter 6. Pseudoscience and the Founding of Eugenics

1. T. Gossett, *Race: The History of an Idea in America* (New York: Schocken Books, 1963), 77–78; and A. Montagu, *Man's Most Dangerous Myth: The Fallacy of Race*, 5th ed. (New York: Oxford University Press, 1974), 107, 314–318.

2. M. Krings, A. Stone, and S. Paabo, "Neandertal DNA Sequences and the Origin of Modern Humans," *Cell* 90, no. 1 (July 11, 1997): 19–30. The authors of this article extracted mitochondrial DNA (MtDNA) from a Neanderthal fossil and found that it was outside of the existing MtDNA diversity in extant humans. This suggests that the Neanderthals did not leave any descendant lineages in modern humans. This interpretation is controversial; see, for example, K. Wong, "Ancestral Quandary," *Scientific American,* January 1998. The controversy stems from the fact that there are genes in modern humans that cannot be found in Africans (indicating that some archaic human genes must have been passed on). What is clear, however, is that the "big-brained" Neanderthals do not represent a main line of human evolution.

3. W. Ndoro, "Great Zimbabwe," *Scientific American,* November 1997, 94–99; see also R. Klein, "The Archeology of Modern Human Origins," *Evolutionary Anthropology* 1, no. 1 (1992): 10–11, for a description of archaeological biases against African civilizations.

4. J. Diamond, "The Accidental Conqueror," *Discover* 10, no. 12 (1989); *The Third Chimpanzee* (New York: HarperCollins, 1992), 235–248; and *Guns, Germs, and Steel: The Fates of Human Societies* (New York: W. W. Norton, 1997).

5. This quotation is from F. Galton, *Hereditary Genius: An Inquiry into Its*

Laws and Consequences (London: Macmillan, 1892), 346. See also A. Chase, *The Legacy of Malthus: The Social Costs of the New Scientific Racism* (New York: Knopf, 1977), 104.

6. The effectiveness of selection on fruit fly life-span is shown in J. L. Graves, E. Toclson, C. M. Jeans, L. U. Vu, and M. R. Rose, "Desiccation Resistance, Flight Duration, Glycogen, and Postponed Senescence in *Drosophila melanogaster*," *Physiological Zoology* 65, no. 2 (1992): 268–286.

7. P. J. Byard, "Quantitative Genetics of Human Skin Color," *Yearbook of Physical Anthropology* 24 (1981): 123–137; and B. K. Rana, E. D. Hewett, L. Jin, B. H. Chang, N. Sambuushin, M. Lin, S. Watkins, M. Bumshad, L. Jorde, T. Jenkins, and W. H. Li, "High Polymorphism at the Human Melanocortin 1 Receptor Locus," *Genetics* 151 (April 1999): 1547–1557. K. Owens and M. King, "Genomic Views of Human History," *Science* (1999): 451–453, suggest that one major locus may control variation of human skin color.

8. Galton, *Hereditary Genius*, 339.

9. Alexis de Tocqueville, "Some Considerations concerning the Present State and Probable Future of the Three Races That Inhabit the Territory of the United States," in *Democracy in America* (Chicago: Encyclopaedia Britannica, 1830), 179.

10. F. Galton, "Hereditary Improvement," *Fraser's Magazine* (1873).

11. L. Terman, *The Measurement of Intelligence: An Explanation of and a Complete Guide for the Use of the Stanford Revision and Extension of the Binet-Simon Intelligence Scale* (Boston: Houghton Mifflin, 1916), 91–92.

12. W. Wilson, *History of the American People* (New York, 1902), 5:212–214, quotation on p. 212.

13. F. Walker, "Restriction of Immigration," *Atlantic Monthly* 77 (June 1896): 828.

14. L. Salyer, *Laws as Harsh as Tigers: Chinese Immigrants and the Shaping of Modern Immigration Law* (Chapel Hill: University of North Carolina Press, 1995), 67–68.

15. *People v. Hall*, 4 Cal. 399 (1854).

Chapter 7. Mendelism, the Neo-Darwinian Synthesis, and the Growth of Eugenics

1. The general rule for quantitative traits is that the number of phenotypic classes equals $2n + 1$, where n is number of genes responsible for the trait.

2. We can use the binomial equation $1.0 = (a + b + c + \ldots + n)^2$ to predict the frequencies of genotypes at this locus. To accomplish this, we must use three variables (a, b, and c) because there are three alleles. We thus arrive at the expression $(a + b + c)^2 = a^2 + 2ab + 2ac + 2bc + b^2 + c^2 = 1.0$, and the six terms in the equation correspond to the six possible genotypes. If we allow the frequency of blood group $A = a = 0.334$, $B = b = 0.333$, and $O = c = 0.333$, then the equation for the genotype frequencies is $(AA) + 2(AB) + 2(AO) + 2(BO) + (BB) + (OO)$; and the frequencies of the blood group are $0.111 + 0.223 + 0.223 + 0.223 + 0.111 + 0.111 = 1.0$.

3. F. Galton, *Hereditary Genius: An Inquiry into Its Laws and Consequences* (London: Macmillan, 1892), 412.

4. R. C. Punnett, "Eliminating Feeblemindedness," *Journal of Heredity* 8 (1917): 464–465.

5. A. Chase, *The Legacy of Malthus: The Social Costs of the New Scientific Racism* (New York: Knopf, 1977), 111.

6. Ibid., 134.

7. Ibid., 313–318; S. J. Gould, *The Mismeasure of Man* (New York: Norton, 1981), 335; and P. Reilly, *The Surgical Solution: A History of Involuntary Sterilization in the United States* (Baltimore: Johns Hopkins University Press, 1991), 86–87.

8. H. H. Laughlin, *The Legal Status of Eugenical Sterilization: History and Analysis of Litigation under the Virginia Sterilization Statute* (Chicago: Fred J. Ringley, 1930), 16–19.

9. [J. Stiles], *Popular Science Monthly,* February 1903, 382–383, described in Chase, *The Legacy of Malthus,* 197.

10. Chase, *The Legacy of Malthus,* 210.

11. Ibid., 224.

12. C. B. Davenport and M. Steggerda, *Race Crossing in Jamaica* (Washington, D.C.: Carnegie Institution of Washington, 1929).

13. Ibid., 469.

14. E. Barkan, *The Retreat of Scientific Racism: Changing Concepts of Race in Britain and the United States between the World Wars* (New York: Cambridge University Press, 1992), 163.

15. Ibid., 67–68.

16. Chase, *The Legacy of Malthus,* 116.

17. H. J. Muller, *Out of the Night: A Biologist's View of the Future* (New York: Vanguard Press, 1935).

Chapter 8. Eugenics, Race, and Fascism

1. D. Gasman, *The Scientific Origins of National Socialism: Social Darwinism in Ernst Haeckel and the German Monist League* (New York: American Elsevier Press, 1971), 157.

2. Ibid., 147–182.

3. B. Müller-Hill, *Murderous Science: Elimination by Scientific Selection of Jews, Gypsies, and Others in Germany, 1933–1945* (Oxford: Oxford University Press, 1988), 7–8.

4. Ibid., 9.

5. S. Kühl, *The Nazi Connection: Eugenics, American Racism, and German National Socialism* (Oxford: Oxford University Press, 1994), 27.

6. Ibid.

7. Ibid., 37.

8. A. Chase, *The Legacy of Malthus: The Social Costs of the New Scientific Racism* (New York: Knopf, 1977), 348.

9. Kühl, *The Nazi Connection,* 101–102; and M. Grodin and G. Annus,

The Nazi Doctors and the Nuremberg Code (Oxford: Oxford University Press, 1992).

10. Kühl, *The Nazi Connection,* 352.

11. Müller-Hill, *Murderous Science,* 49–50.

12. W. L. Shirer, *The Rise and Fall of the Third Reich: A History of Nazi Germany* (Greenwich, Conn.: Fawcett Crest Books, 1959), 1224.

13. Müller-Hill, *Murderous Science,* 12.

14. Estimated population losses in World War II are given in J. Noakes and G. Pridham, *Nazism 1919–1945,* Vol. 3, *Foreign Policy, War, and Racial Extermination* (Exeter, U.K.: Exeter University Press, 1988), 874.

15. Ibid., 1208.

Chapter 9. The Retreat and Postwar Revival of Scientific Racism

1. This pamphlet was banned because of the activity of Kentucky representative Andrew J. May, then chair of the House Military Affairs Committee. Predictably, after the war May would be convicted for bribery and conspiracy while in office; this affair was reported in 1944 by G. Reynolds, "What the Negro Soldier Thinks of This War," in *A Documentary History of the Negro People,* ed. H. Aptheker (Secaucus, N.J.: The Citadel Press, 1974), 4:491.

2. Hitler arrived in the stadium with Max Schmeling and some other celebrities in time to see another German victory in the hammer throw event. During the medal ceremony Hitler acknowledged Owens, but he did not greet Owens in his box. It should be stated that Hitler had been asked the previous day by the Olympic Committee not to greet victorious athletes in his box. When approached by his own publicists about the possibility of having a photograph taken with Jesse Owens, he violently replied: "Do you really think that I will allow myself to be photographed shaking the hands of a negro?" (D. Hart-Davis, *Hitler's Games: The 1936 Olympics* [London: Century Press, 1986], 176–177).

3. L. C. Dunn and Th. Dobzhansky, *Heredity, Race, and Society* (New York: Mentor Books, 1946), 110.

4. See for example, L. L. Cavalli-Sforza, P. Menozzi, and A. Piazza, *The History and Geography of Human Genes* (Princeton, N.J.: Princeton University Press, 1994); and A. Templeton, "The Genetic and Evolutionary Significance of Human Races," in *Understanding Race and Intelligence,* ed. J. Fish (Northvale, N.J.: Jason Aronson, in press).

5. A. S. Wiener, "Blood Group Factors and Racial Relationships," *Science* 103 (1946): 147; and J. Marks, "The Legacy of Serological Studies in American Physical Anthropology," *History and Philosophy of the Life Sciences* 18 (1996): 352.

6. UNESCO, *Race and Science: The Race Question in Modern Science* (New York: Columbia University Press, 1961), 493.

7. Ibid., 496.

8. Lenz is quoted in UNESCO, *The Race Concept: Results of an Inquiry* (Paris: UNESCO, 1952), 30–31.

9. UNESCO, *Race and Science,* 502.

10. The 1950 statement was drafted by a committee made up of Ernest Beaglehole (New Zealand), Juan Comas (Mexico), L. A. Costa Pinto (Brazil), E. Franklin Frazier (U.S.), Morris Ginsburg (U.K.), Humayun Kabir (India), Claude Lévi-Strauss (France), and M. F. Ashley-Montagu (U.S.). The list of notable geneticists who reviewed the statement before its publication includes E. G. Conklin, Th. Dobzhansky, L. C. Dunn, Julian S. Huxley, and H. J. Muller. In addition the statement was reviewed by a number of other scientists and sociologists, such as Otto Klineberg, Gunnar Myrdal (author of *An American Dilemma*), and Curt Stern.

The members of the 1951 committee were Professor R.A.M. Bergman, Royal Tropical Institute, Amsterdam; Professor Gunnar Dahlberg, Director, State Institute for Human Genetics and Race Biology, University of Uppsala; Professor L. C. Dunn, Department of Zoology, Columbia University, New York; Professor J.B.S. Haldane, Head, Department of Biometry, University College, London; Professor M. F. Ashley Montagu, Chairman, Department of Anthropology, Rutgers University, New Brunswick, N.J.; Dr. A. E. Mourant, Director, Blood Group Reference Laboratory, Lister Institute, London; Professor Hans Nachtscheim, Director, Institut für Genetik, Freie Universität, Berlin; Dr. Eugène Schreider, Directeur Adjoint du Laboratoire d'Anthropologie Physique de l'École des Hautes Études, Paris; Professor Harry L. Shapiro, Chairman, Department of Anthropology, American Museum of Natural History, New York; Dr. J. C. Trevor, Faculty of Archaeology and Anthropology, University of Cambridge; Dr. Henri V. Vallois, Professeur au Muséum d'Histoire Naturelle, Directeur du Musée de l'Homme, Paris; Professor S. Zuckerman, Head, Department of Anatomy, Medical School, University of Birmingham; Professor Th. Dobzhansky, Department of Zoology, Columbia University, New York. Dr. Julian Huxley contributed to the final wording. The text was drafted, at UNESCO House, Paris, on June 8, 1951. This list of the committee members comes from UNESCO, *Race and Science*, 506.

11. G. Myrdal, *An American Dilemma: The Negro Problem and Modern Democracy*, with the assistance of Richard Sterner and Arnold Rose (New York: Harper and Brothers, 1944), 339.

12. R. Kluger, *Simple Justice: The History of Brown v. Board of Education and Black America's Struggle for Equality* (New York: Vintage Books, 1977), 257.

13. H. Garrett, Letter to the Editor, *Science* 135 (1962): 984. See also W. H. Tucker, *The Science and Politics of Racial Research* (Urbana: University of Illinois Press, 1994), 154.

Part Four. Biological Theories of Race at the Millennium

1. M. Nei and A. K. Roychoudhury, "Genetic Relationship and Evolution of Human Races," *Evolutionary Biology* 14 (1982): 1–59, quotation on p. 11.

2. "No Such Thing as Race, Genetic Studies Say," *Arizona Republic*, February 21, 1995, A4. This article also appeared in the *Los Angeles Times, San Jose Mercury News,* and other newspapers.

3. The 1995 meeting of the AAAS was held in Atlanta and included a panel entitled Is Race a Legitimate Concept for Science? This panel included the following participants: Dr. Joseph L. Graves Jr. (Arizona State University; added after program announcement was published), Jane Maienshien (Arizona State University; did not speak because of illness), John Ladd (Brown University), Roscoe Lee Brown (Brown University), Michael Omi (University of California, Berkeley), Evelyn Hu DeHart (University of Colorado–Boulder), C. Loring Brace (University of Michigan), and Solomon Katz (University of Pennsylvania, president of the American Association of Physical Anthropologists).

4. Kidd is quoted in E. Marshall, "DNA Studies Challenge the Meaning of Race," *Science* 282 (1998): 654–655, quotation on p. 655.

Chapter 10. The Race and IQ Fallacy

1. A. Chase, *The Legacy of Malthus: The Social Costs of the New Scientific Racism* (New York: Knopf, 1977), 318–321.

2. Ibid., 466.

3. A. Jensen, "How Much Can We Boost IQ and Scholastic Achievement?" *Harvard Educational Review* 39 (1969): 1.

4. Chase, *The Legacy of Malthus*, 466; O. Walker, "The Windsor Hills Story," *Integrated Education* 8, no. 3 (May–June 1970): 4–9.

5. Ehrlichman describes the Nixon agenda in *Witness to Power: The Nixon Years* (New York: Simon and Schuster, 1982), 133. The Moynihan quotation comes from J. Neary, "A Scientist's Variation on a Disturbing Racial Theme," *Life,* June 12, 1970, 58d. See also W. H. Tucker, *The Science and Politics of Racial Research* (Urbana: University of Illinois Press, 1994), 207.

6. Frances Lawrence, address to his faculty, reported by the Associated Press, January 31, 1995.

7. R. J. Herrnstein and C. Murray, *The Bell Curve: Intelligence and Class Structure in American Life* (New York: Free Press, 1994), 277.

8. The data on intelligence differences between blacks and whites were summarized from A. Shuey, *The Testing of Negro Intelligence,* 2d ed. (New York: Social Science Press, 1966); R. T. Osborne and F. McGurk, eds., *The Testing of Negro Intelligence,* vol. 2 (Athens, Ga.: Foundation for Human Understanding, 1982); J. Sattler, *Assessment of Children's Intelligence and Other Special Abilities,* 2d ed. (Boston: Allyn and Bacon, 1988); K. R. Vincent, "Black/White IQ Differences: Does Age Make a Difference?" *Journal of Clinical Psychology* 47 (1991): 266–270; A. R. Jensen, "The Nature of Black-White Difference on Various Psychometric Tests: Spearman's Hypothesis," *The Behavioral and Brain Sciences* 8 (1985): 193–258; and A. R. Jensen, "Psychometric g and Achievement," in *Policy Perspectives on Educational Testing,* ed. B. R. Gifford (Boston: Kluwer Academic Publishers, 1993), 117–227.

9. M. R. Rose, *Evolutionary Biology of Aging* (New York: Oxford University Press, 1991), 33.

10. J. R. Flynn, "The Mean IQ of Americans: Massive Gains 1932 to 1978," *Psychological Bulletin* 101 (1984): 171–191.

11. B. Devlin, M. Daniels, and K. Roeder, "The Heritability of IQ," *Nature* 388, no. 6641 (1997): 368-371.

12. Herrnstein and Murray, *The Bell Curve,* 308.

13. C. Y. Valenzuela, C. S. Pastene, and C. M. Perez, "Intelligence and Genetic Markers in Chilean Children," *Biological Research* 31, no. 2 (1998): 81-92.

14. A. K. Roychoudhury and M. Nei, *Human Polymorphic Genes: World Distribution* (New York: Oxford University Press, 1988), 185-188.

15. J. L. Graves, "The Misuse of Life History Theory: J. P. Rushton and the Pseudoscience of Racial Hierarchy," in *Understanding Race and Intelligence,* ed. J. Fish (Northvale, N.J.: Jason Aronson, in press).

16. J. Kozol, *Savage Inequalities: Children in America's Schools* (New York: Harper Perennial, 1991). R. T. Schaefer, *Racial and Ethnic Groups,* 7th ed. (New York: Longmans, 1998), 22-23.

17. B. Bryant and P. Mohai, eds., *Race and the Incidence of Environmental Hazards: A Time for Discourse* (Boulder, Colo.: Westview Press, 1992), 15-16.

Chapter 11. The Race and Disease Fallacy

1. C. P. Jones, T. A. LaViest, and M. Lillie-Blanton, "'Race' in the Epidemiologic Literature: An Examination of the *American Journal of Epidemiology,* 1921-1990," *American Journal of Epidemiology* 134 (1991): 1079-1084.

2. These figures are calculated by utilizing male and female incidence data from U. S. Census Bureau, *Statistical Abstract of the United States* (Washington, D.C.: GPO, 1998), 102.

3. A. P. Polednak, *Racial and Ethnic Differences in Disease* (New York: Oxford University Press, 1989), 71.

4. R. A. Irvine, M. Yu, R. Ross, and G. Coetzee, "The CAG and GGC Microsatellites of the Androgen Receptor Gene Are in Linkage Disequilbrium in Men with Prostate Cancer," *Cancer Research* 55 (1995): 1937-1940.

5. O. Sartor, Q. Zheng, and J. Eastham, "Androgen Receptor Gene CAG Repeat Length Varies in a Race-Specific Fashion in Men without Prostate Cancer," *Urology* 53, no. 2 (1999): 378-380; and L. Correa-Cerro, G. Wohr, J. Haussler, P. Berthon, E. Drelon, P. Mungin, G. Fournier, O. Cussenot, P. Kraus, W. Just, T. Paiss, J. M. Cantu, and W. Vogel, "(CAG)nCAA and GGN Repeats in the Human Androgen Receptor Gene Are Not Associated with Prostate Cancer in a French-German Population," *European Journal of Human Genetics* 7, no. 3 (1999): 357-362.

6. J. Peto, N. Collins, R. Barfoot, S. Seal, W. Warren, N. Rahman, D. Easton, C. Evans, J. Deacon, and R. Stratton, "Prevalence of BRCA1 and BRCA2 Gene Mutations in Patients with Early-Onset Breast Cancer," *Journal of the National Cancer Institute Bethesda* 91, no. 11 (June 2, 1999): 943-949.

7. J. Caldwell and P. Caldwell, "The African AIDS Epidemic," *Scientific American,* March 1996, 62-69.

8. J. Rushton, *Race, Evolution, and Behavior: A Life History Perspective* (New Brunswick, N.J.: Transaction Publishers, 1995), 178-183.

9. J. Jones, *Bad Blood: The Tuskegee Syphilis Experiment* (New York: The Free Press, 1981), 27.

10. M. A. Haynes and B. Smedley, eds., *The Unequal Burden of Cancer: An Assessment of NIH Research and Programs for Ethnic Minorities and the Medically Underserved* (Washington, D.C.: National Academy Press, 1999), 71–73 and figure 2.1.

Conclusion. What Can or Will We Do without Race?

1. G. Langer, "Prayers for Peace: Racism, War Concerns in the Coming Millennium," http://abcnews.go.com/ABC2000/DailyNews/abcpoll991004.html, October 4, 1999.

2. "Double-Checks to Count as Minorities in Census," *Arizona Republic,* March 11, 2000, A3.

3. J. Ladd, "Philosophical Reflections on Race and Racism," *American Behavioral Scientist* 41, no. 2 (1997): 212–222.

4. C. M. Steele and J. Aronson, "Stereotype Threat and the Intellectual Test Performance of African Americans," *Journal of Personality and Social Psychology* 69, no. 5 (1995): 797–781; and M. Lovaglia, J. W. Lucas, J. A. Houser, S. Thye, and B. Markovsky, "Status Processes and Mental Ability Test Scores," *American Journal of Sociology* 104, no. 1 (1998): 195–228.

5. D. L. Kigar, S. F. Witelson, I. Glezer, and T. Harvey, "Estimates of Cell Number in Temporal Neocortex in the Brain of Albert Einstein," *Society for Neuroscience Abstracts* 23, nos. 1–2 (1997): 213.

Appendix A

1. A. K. Roychoudhury and M. Nei, *Human Polymorphic Genes: World Distribution* (Oxford: Oxford University Press, 1988), 236, table 170. See also the most recent frequencies at the National HLA Registry: http://www.swmed.edu/home_pages/ASHI/prepr/mori_gf.htm.

2. D. Hartl and A. G. Clark, *Principles of Population Genetics*, 2d ed. (Sunderland, Mass.: Sinauer Associates, 1989), 308–309.

Bibliography

Chapter 1. The Earliest Theories

By necessity I rely on a number of works to build the historical narrative in part 1. In particular, the following studies of the formation of the Western race concept were indispensable: Thomas Gossett, *Race: The History of an Idea in America* (New York: Schocken Books, 1963); Winthrop Jordan, *White over Black: American Attitudes toward the Negro, 1550–1812* (Chapel Hill: University of North Carolina Press, 1968); UNESCO, *Race and Science: The Race Question in Modern Science* (New York: Columbia University Press, 1961); Ashley Montagu, *Man's Most Dangerous Myth: The Fallacy of Race,* 5th ed. (New York: Oxford University Press, 1974); John R. Baker, *Race* (New York: Oxford University Press, 1974); Audrey Smedley, *Race in North America: Origin and Evolution of a World View* (Boulder, Colo.: Westview Press, 1993); J. Philippe Rushton, *Race, Evolution, and Behavior: A Life History Perspective* (New Brunswick, N.J.: Transaction Publishers, 1995); and Hannah Augstein, ed., *Race: The Origins of an Idea, 1760–1850* (Chippenham, U.K.: Anthony Rowe, 1996). This is by no means an all-inclusive list, but it represents the spectrum of perspectives on the development of early racial theory in the Western world.

My historical analysis of slavery was supported by Kenneth Stampp, *The Peculiar Institution: Slavery in the Antebellum South* (Toronto: Vintage, 1963); and Keith R. Bradley, *Slavery and Rebellion in the Roman World, 140 B.C.–70 B.C.* (Bloomington: University of Indiana Press, 1989); Vincent Harding, *There Is a River* (New York: Vintage Press, 1981); and many other works discussing the depravity of American slavery.

The history of anti-Semitism in Europe is well examined in Hugo Valentin, *Anti-Semitism Historically and Critically Examined,* trans. A. G. Chater (1936; reprint, Freeport, N.Y.: Books for Libraries, 1971); Franklin Young, *Understanding the New Testament* (Englewood Cliffs, N.J.: Prentice-Hall, 1957); and B. Lazare, *Anti-Semitism: Its Roots and Historical Causes* (London: Britons Publishing, 1967).

The writings of Saint Augustine are also referenced (circa A.D. 412); see *The City of God,* translated by Marcus Dods, with an introduction by Thomas Merton (New York: Modern Library, 1950).

Chapter 2. Colonialism, Slavery, and Race in the New World

My analysis of the impact of the Atlantic slave trade on international economic developments is based on the following works: Eric Williams, *Capitalism and Slavery* (Chapel Hill: University of North Carolina Press, 1944); Elizabeth Donnan, ed., *Documents Illustrative of the History of the Slave Trade to America* (New York: Octagon Books, 1965); Philip D. Curtin, *The Atlantic Slave Trade: A Census* (Madison: University of Wisconsin Press, 1969); John D. Fage, *A History of Africa* (London: Hutchinson, 1978); Colin Palmer, *Human Cargoes: The British Slave Trade to Spanish America, 1700–1739* (Urbana: University of Illinois Press, 1981); J. Devisse and S. Labib, "Africa in Inter-Continental Relations," in *General History of Africa*, Vol. 4, *Africa from the Twelfth to the Sixteenth Century*, ed. D. T. Niane (London: Heinemann; Berkeley: University of California Press, 1984); and Joseph E. Inikori and Stanley L. Engerman, eds., *The Atlantic Slave Trade: Effects on Economies, Societies, and Peoples in Africa, the Americas, and Europe* (Durham, N.C.: Duke University Press, 1992).

Helpful additional references for the development of Spanish colonialism in the New World are Roger Merriman, *The Rise of the Spanish Empire in the Old World and in the New*, Vol. 2, *The Catholic Kings* (New York: Cooper Square, 1962); and Bartolomé de Las Casas, *History of the Indies*, trans. and ed. Andrée Collard (New York: Harper and Row, 1971). The inaccuracies in Las Casas's estimates are treated in Lewis Hanke, *Bartolomé de Las Casas: Bookman, Scholar, and Propagandist* (Philadelphia: University of Pennsylvania Press, 1952); and Douglas Ubelaker, "North American Indian Population Size," in *Disease and Demography in the Americas*, ed. John W. Verano and Douglas Ubelaker (Washington, D.C.: Smithsonian Institution Press, 1992).

Other references for this chapter include Samir Amin, *Neo-Colonialism in West Africa;* trans. Francis McDonagh (New York: Monthly Review Press, 1973); Maris Vinovskis, *Demographic History and the World Population Crisis* (Worcester, Mass.: Clark University Press, 1976); Robert Rotberg and Theodore K. Rabb, eds., *Population and Economy: Population and History from the Traditional to the Modern World* (Cambridge: Cambridge University Press, 1986); United Nations Department of International Economic Studies, *Fertility Behavior in the Context of Development: Evidence from the World Fertility Study* (New York: United Nations, 1987); and Robert Fogel and Stanley L. Engerman, *Time on the Cross: The Economics of American Negro Slavery* (1974; reprint, Lanham, Md.: University Press of America, 1984).

The sections on admixture are supported by Winthrop Jordan, *White over Black: American Attitudes toward the Negro, 1550–1812* (Chapel Hill: University of North Carolina Press, 1968), 167–178; Daniel Hartl and Andrew G. Clark, *Principles of Population Genetics*, 2d ed. (Sunderland, Mass.: Sinauer Associates, 1989); Annette Gordon-Reed, *Thomas Jefferson and Sally Hemings: An American Controversy* (Charlottesville: University Press of Virginia, 1997); E. Foster, M. A Joblina, and P. G. Taylor, "Jefferson Fathered Slave's Last Child," *Nature* 396, no. 6706 (1998): 27–28; Edward T. Reed, "Caucasian Genes in American Negroes," *Science* 165 (1969): 762–788; L. Luca Cavalli-

Sforza and William F. Bodmer, *The Genetics of Human Populations* (San Francisco: W. H. Freeman, 1971); and finally L. Luca Cavalli-Sforza, Paolo Menozzi, and Alberto Piazza, *The History and Geography of Human Genes* (Princeton, N.J.: Princeton University Press, 1994).

Chapter 3. Pre-Darwinian Theories of Biology and Race

In addition to the general summaries of racial thought listed for chapter 1, the following source was consulted: Ivan Hannaford, *Race: The History of an Idea in the West* (Baltimore: Johns Hopkins University Press, 1996). The original title of François Bernier's 1684 essay on the classification of the human races was "Nouvelle division de la terre par les differents espèces ou races qui l'habitent . . . ," John Ray, *The Wisdom of God Manifested in the Works of Creation* (London: Printed for Samuel Smith . . . , 1692).

Toward the end of the seventeenth century, Gottfried Leibniz published comments on Bernier's 1684 essay (*Otium Hanoveranum siue miscellanes ex ore* . . . [Leipzig, 1718], 37). See also Henry Home, Lord Kames, "Preliminary Discourse," in *Race: The Origins of an Idea, 1760–1850,* ed. Hannah Augstein (Chippenham, U.K.: Anthony Rowe, 1996); anonymous review of Charles White's *An Account of Regular Gradation in Man* (1799), *Monthly Review,* n.s., 33 (1800): 360–364; anonymous review of James Prichard's *The Natural History of Man* (1842), *Eclectic Review,* 12 (1848); 660–666; Samuel Stanhope Smith, *An Essay on the Causes of the Variety of Complexion and Figure in the Human Species,* ed. Winthrop D. Jordan (Cambridge, Mass.: Belknap Press of Harvard University Press, 1965); Johann Friedrich Blumenbach, *The Anthropological Treatises of Johann Friedrich Blumenbach,* translated from the Latin, German, and French originals by Thomas Bendyshe (London: Longman, Green, Roberts, and Green, 1865), which includes the first (1775) and third (1857) editions of the author's *On the Natural Variety of Mankind;* and Josiah C. Nott and George R. Gliddon, *Indigenous Races of the Earth; or, New Chapters of Ethnological Inquiry* (Philadelphia: J. P. Lippincott, 1857), ethnographic tables.

My views on Jefferson are supported by Winthrop Jordan, *White over Black: American Attitudes toward the Negro, 1550–1812* (Chapel Hill: University of North Carolina Press, 1968); and Max Beloff, *Thomas Jefferson and American Democracy* (New York: Macmillan, 1949), 429–481, which is a detailed analysis of Jefferson's views on race, his slaveholding, personal contradictions, and democracy. These views were all originally printed in Thomas Jefferson's *Notes on the State of Virginia* (Philadelphia: H. C. Carey and I. Lea . . . , 1825).

Also cited are Frederick Douglass, "The Claims of the Negro Ethnologically Considered," in *The Life and Writings of Frederick Douglass,* Vol. 2, *Pre–Civil War Decade,* ed. Phillip Foner (New York: International Publishers, 1952). Herbert Aptheker, *American Negro Slave Revolts* (New York: International Publishers, 1943). Hinton Rowan Helper, *The Impending Crisis of the South and How to Meet It,* ed. George M. Fredrickson (1857; reprint, with a new introduction by the editor, Cambridge: Belknap Press of Harvard University Press, 1968).

For a readable account of why and how evolutionary theory organizes research in biology, see Michael R. Rose, *Darwin's Specter* (Princeton, N.J.: Princeton University Press, 1999). The history of evolutionary reasoning is supplied by Ernst Mayr, *The Growth of Biological Thought: Diversity, Evolution, and Inheritance* (Cambridge: Belknap Press of Harvard University Press, 1984).

Chapter 4. Darwinism Revolutionizes Anthropology

The historical material on the life of Charles Darwin and on his attitudes toward race and slavery is found in Adrian Desmond and James Moore, *Darwin* (London: Michael Joseph, 1991); *Charles Darwin's Notebooks, 1836–1844*, transcribed and edited by Paul H. Barrett, Sandra Herbert, Sydney Smith, Peter Gautrey, and David Kohn (New York: Cambridge University Press, 1987); P. Manning, *Slavery and African Life: Occidental, Oriental, and African Slave Trades* (Cambridge: Cambridge University Press, 1990).

The X Club and the ASL debates are described in P. Fryer, *Staying Power: The History of Black People in Britain* (London: Pluto Press, 1984), 176–179; Thomas Gossett, *Race: The History of an Idea in America* (New York: Schocken Books, 1963), 95–97; and Desmond and Moore, *Darwin*, 521–534.

Original writings from the period include James Hunt, "On the Negro's Place in Nature," *Memoirs of the Anthropological Society of London* 1 (1863–64): 51–52; James Hunt, "Address Delivered at the Third Anniversary Meeting of the Anthropological Society of London," *Journal of the Anthropological Society of London* 4 (1866): lxxviii; Commander Bedford Pimm, *The Negro in Jamaica* (London: Trubner, 1866), 15, 16, 35, 50, 51, 63.

The 1755 essay of Thomas R. Malthus is *An Essay on the Principle of Population, or, A View of Its Past and Present Effects on Human Happiness: With an Inquiry into Our Prospects respecting the Future Removal or Mitigation of the Evils Which It Occasions* (Cambridge: Cambridge University Press, 1988). Malthus's views in turn may be compared to the modern demographic analysis in W. L. Burn, "The Population of Ireland, 1750–1845," *The Economic History Review*, 2d ser., 4 (1951–1952): 256–257; and H. Moller, ed., *Population Movements in Modern European History* (New York: Macmillan, 1964).

The importance of evolutionary biology has been well discussed; see, for example, Ernst Mayr, *The Growth of Biological Thought: Diversity Evolution, and Inheritance* (Cambridge: Belknap Press of Harvard University Press, 1984); Richard Dawkins, *The Blind Watchmaker: Why the Evidence of Evolution Reveals a Universe without Design* (New York: W. W. Norton, 1987); Michael R. Rose, *The Evolutionary Biology of Aging* (New York: Oxford University Press, 1991); George C. Williams and Randolph M. Neese, *Why We Get Sick: The New Science of Darwinian Medicine* (New York: Oxford University Press, 1994); Thomas F. Glick, ed., *The Comparative Reception of Darwinism* (Austin: University of Texas Press, 1974); James Moore, *The Post-Darwinian Controversies: A Study of the Protestant Struggle to Come to Terms with Darwin in Great Britain and America, 1870–1900* (Cambridge: Cambridge University Press, 1974).

Good references on the coming of social Darwinism include Loren Eisley, *Darwin's Century: Evolution and the Men Who Discovered It* (New York: Anchor/Doubleday, 1958); and Richard Hofstadter, *Social Darwinism in American Thought*, rev. ed. (New York: George Braziller, 1959).

Chapter 5. Race and Social Darwinism

The following sources provide general background for the history of Reconstruction: W.E.B. Du Bois, *Black Reconstruction* (New York: Atheneum, 1935); Kenneth Stampp, *The Era of Reconstruction, 1865–1877* (New York: Vintage Books, 1974); Thomas Page, *Red Rock: A Chronicle of Reconstruction* (New York: Arden, 1983); James McPherson, *Ordeal by Fire: The Civil War and Reconstruction* (New York: Knopf, 1984); and Richard W. Murphy, *The Nation Reunited: War's Aftermath* (Alexandria, Va.: Time-Life, 1987).

In "The Status of the Negro: 1865–1915," in *Race: The History of an Idea in America* (New York: Schocken Books, 1963), Thomas Gossett describes the racial ideology of this period in detail. Richard Hofstadter, *Social Darwinism in American Thought*, rev. ed. (New York: George Braziller, 1959), is still the classic work on the impact of Spencerism (there is a 1992 edition [Boston: Beacon Press] with an introduction by Eric Foner). Additional general background is supplied by Allan Chase, *The Legacy of Malthus: The Social Costs of the New Scientific Racism* (New York: Knopf, 1977).

Original works from this period of interest are Herbert Spencer, *The Principles of Sociology* (New York: D. Appleton, 1876–1897); William Sumner, *What Social Classes Owe to Each Other* (New York: Harper and Bros., 1883); and Lester Ward, *Pure Sociology* (New York: Macmillan, 1903).

Charles Darwin, *The Expression of Emotions in Animals and Man* (London: Murray, 1872), and Edward O. Wilson, *Sociobiology* (Cambridge: Harvard University Press, 1975), are classic works outlining how evolutionary biologists originally constructed the evolution of society. Marxian views on this subject can be garnered from Karl Marx, *The Poverty of Philosophy* (New York: International Publishers, 1963); K. Marx and Frederich Engels, "The Manifesto of the Communist Party," and Frederich Engels, "The Part Played in Labour for the Transition of Ape to Man," both in *Marx and Engels Collected Works* (Moscow: Progress Publishers, 1969).

Chapter 6. Pseudoscience and the Founding of Eugenics

The story of the origin of the theory of continental drift is recounted in Martin Schwarzbach, *Alfred Wegener: The Father of Continental Drift* (Madison, Wisc.: Science Tech, 1986). General references used in this chapter include John R. Baker, *Race* (New York: Oxford University Press, 1974); and Michael D. Biddiss, *Father of Racist Ideology: The Social and Political Thought of Count Gobineau* (New York: Weybright and Talley, 1970). Michael D. Biddiss, ed., *Gobineau: Selected Political Writings* (New York: Harper and Row, 1970), supplied much of the background on Gobineau. See also Gobineau's seminal 1915 essay, *The Inequality of Human Races,* trans. Adrian Collins, with an introduction by Oscar Levy (New York: H. Fertg, 1967); and Ellsworth Huntington, *The Character of Races: As Influenced by Physical Environment, Natural Selection, and Historical Development* (New York: Scribner and Sons, 1924). Huntington's work presents the theory of cold weather as a selective agent for intelligence in Europeans (47–60). In contrast to Gobineau, see Alexis de Tocqueville's 1830 essay, "Some Considerations concerning the Present State and Probable Future of the Three Races That Inhabit the Territory of the United States," in *Democracy in America* (Chicago: Encyclopaedia Britannica, 1830). Finally, general information on climatic factors in the development of early African societies can be gathered from J. D. Fage and R. Oliver, *The Cambridge History of Africa,* Vol. 3., *From c. 1050 to c. 1600* (New York: Cambridge University Press, 1978).

An excellent essay that outlines modern evolutionary perspectives on the development of mental ability is Th. Dobzhansky and A. Montagu, "Natural Selection and the Mental Capacities of Mankind," *Science* 105 (1947): 587–590. The disease pedigrees of the English royal families can be found in Gordon Edlin, *Genetic Principles: Human and Social Consequences* (Boston: Jones and Bartlett, 1984), 282–284.

Galton's dysgenics formulated in modern terms is found in A. Jensen, "How Much Can We Boost IQ and Scholastic Achievement?" *Harvard Educational Review* 39 (1969); 1–123; and the most recent iteration of this argument is Richard J. Herrnstein and Charles Murray, *The Bell Curve: Intelligence and Class Structure in American Life* (New York: Free Press, 1994), 341–368.

My information on the anti-immigration movement is based on Thomas Gossett, *Race: The History of an Idea in America* (New York: Schocken Books, 1963); Allan Chase, *The Legacy of Malthus: The Social Costs of the New Scientific Racism* (New York: Knopf, 1977); and K. Ludmerer, "Genetics, Eugenics, and the Immigration Restriction Act of 1924," in *American Immigration and Ethnicity: Nativism, Discrimination, and Images of Immigrants,* ed. G. Pozzetta (New York: Garland Publishing, 1991). Additional information on Chinese immigration and labor is based on Eric Foner, *The Black Worker in America* (Moscow: International Publishers, 1974); and Henry Tsai, *The Chinese Experience in America* (Bloomington: Indiana University Press, 1981).

Chapter 7. Mendelism, the Neo-Darwinian Synthesis, and the Growth of Eugenics

The following are invaluable sources on the history of population genetics: William Provine, *The Origins of Theoretical Population Genetics* (Chicago: University of Chicago, 1971); Ernst Mayr, *The Growth of Biological Thought: Diversity, Evolution, and Inheritance* (Cambridge: Belknap Press of Harvard University Press, 1984); P. Bowler, *The Mendelian Revolution: The Emergence of Hereditarian Concepts in Modern Science and Society* (Baltimore: Johns Hopkins University Press); and Vítezslav Orel, *Gregor Mendel: The First Geneticist* (New York: Oxford University Press, 1996).

Eugenics is one of the most written about topics in the history of biology. My treatment of this subject is based on material contained within the following sources: Leslie C. Dunn and Theodosius Dobzhansky, *Heredity, Race, and Society* (New York: Mentor Books, New American Library, 1946); Allan Chase, *The Legacy of Malthus: The Social Costs of the New Scientific Racism* (New York: Knopf, 1977); G. Allen, "The Eugenics Record Office at Cold Spring Harbor, 1910–1940," *Osiris* 2 (1986):225–264; Mark Adams, ed., *The Wellborn Science: Eugenics in Germany, France, Brazil, and Russia* (New York: Oxford University Press, 1990); Elazar Barkan, *The Retreat of Scientific Racism: Changing Concepts of Race in Britain and the United States between the World Wars* (Cambridge: Cambridge University Press, 1992); Stefan Kühl, *The Nazi Connection: Eugenics, American Racism, and German National Socialism* (Oxford: Oxford University Press, 1994); Edward Larson, *Sex, Race, and Science: Eugenics in the Deep South* (Baltimore: Johns Hopkins University Press, 1995); and Dorothy Nelkin and M. Susan Lindee, *The DNA Mystique: The Gene as a Cultural Icon* (New York: W. H. Freeman, 1995).

Original works examined in this chapter included H. S. Jennings, "Heredity, Variation, and Evolution in Protozoa: Heredity and Variation of Size and Form in *Paramecium*, with Studies of Growth, Environmental Action, and Selection," *Proceedings of the American Philosophical Society* 47 (1908): 393–546; Charles B. Davenport and Morris Steggerda, *Race Crossing in Jamaica* (Washington, D.C.: Carnegie Institution of Washington, 1929); Gertrude C. Davenport and C. B. Davenport, "Heredity of Hair Form in Man," *American Naturalist* 42 (1908): 341–349; C. B. Davenport, "Heredity of Some Human Physical Characteristics," *Proceedings of the Society for Experimental Biology and Medicine* 5 (1908): 101–102; C. B. Davenport, "Heredity of Skin Pigmentation in Man," *American Naturalist* 44 (1910): 642–672; Herman J. Muller, *Out of the Night: A Biologist's View of the Future* (New York: Vanguard Press, 1935); and J.B.S. Haldane, *Heredity and Politics* (London: Allen and Unwin, 1938).

General African American history in this period can be found in Alton Hornsby, *Chronology of African-American History: Significant Events and People from 1619 to the Present* (Detroit: Gale Research, 1991), 71–79; and the story of the town of Rosewood, Florida, is told in Michael D'Orso, *Like Judgement Day: The Ruin and Redemption of a Town Called Rosewood* (New York: G. P. Putnam's Sons, 1996).

Chapter 8. Eugenics, Race, and Fascism

The following sources are useful in understanding the development of Nazi racial ideology and the race hygiene movement: Leslie C. Dunn and Theodosius Dobzhansky, *Heredity, Race, and Society* (New York: Mentor Books, New American Library, 1946); Daniel Gasman, *The Scientific Origins of National Socialism: Social Darwinism in Ernst Haeckel and the German Monist League* (New York: American Elsevier Press, 1971); Allan Chase, *The Legacy of Malthus: The Social Costs of the New Scientific Racism* (New York: Knopf,

1977); Robert J. Lifton, *The Nazi Doctors: Medical Killing and the Psychology of Genocide* (New York: Basic Books, 1986); Benno Müller-Hill, *Murderous Science: Elimination by Scientific Selection of Jews, Gypsies, and Others in Germany, 1933–1945* (Oxford: Oxford University Press, 1988); Michael Kater, *Doctors under Hitler* (Chapel Hill: University of North Carolina Press, 1989); S. F. Weiss, "The Race Hygiene Movement in Germany," in *The Wellborn Science: Eugenics in Germany, France, Brazil, and Russia*, ed. Mark B. Adams (New York: Oxford University Press, 1990); and Stefan Kühl, *The Nazi Connection: Eugenics, American Racism, and German National Socialism* (New York: Oxford University Press, 1994).

Additional useful historical information was gathered from William L. Shirer, *The Rise and Fall of the Third Reich: A History of Nazi Germany* (Greenwich, Conn.: Fawcett Crest Books, 1959). Estimated population losses in World War II are given in Jeremy Noakes and Geoffrey Pridham, *Nazism 1919–1945*, Vol. 3, *Foreign Policy, War, and Racial Extermination* (Exeter, U.K.: Exeter University Press, 1988).

Chapter 9. The Retreat and Postwar Revival of Scientific Racism

General historical references include Gunnar Myrdal, *An American Dilemma: The Negro Problem and Modern Democracy*, with the assistance of Richard Sterner and Arnold Rose (New York: Harper and Brothers, 1944); Gar Alperovitz, *Atomic Diplomacy: Hiroshima and Potsdam: The Use of the Atomic Bomb and the American Confrontation with Soviet Power*, expanded and updated edition (New York: Penguin Books, 1985); Lou Potter, William Miles, and Nina Rosenblum, *Liberators: Fighting on Two Fronts in World War II* (New York: Harcourt Brace Jovanovich, 1991); J. Harvie Wilkinson III, *From Brown to Bakke: The Supreme Court and School Integration, 1954–1978* (Oxford: Oxford University Press, 1979); Richard Kluger, *Simple Justice: The History of Brown v. Board of Education and Black America's Struggle for Equality* (New York: Vintage Books, 1977); Abraham Davis and Barbara Graham, *The Supreme Court, Race, and Civil Rights* (Thousand Oaks, Calif.: Sage Press, 1995).

Material concerning the role of racial research and civil rights is supported by Elazar Barkan, *Retreat of Scientific Racism: Changing Concepts of Race in Britain and the United States between the World Wars* (Cambridge: Cambridge University Press, 1972); William H. Tucker, *The Science and Politics of Racial Research* (Urbana: University of Illinois Press, 1994); Keith Wailoo, *Drawing Blood: Technology and Disease Identity in Twentieth-Century America* (Baltimore: Johns Hopkins University Press, 1997); C. Lane, "The Tainted Sources of *The Bell Curve*," *The New York Review of Books* (December 1, 1994): 4–19; and J. Mercer, "A Fascination with Genetics: Pioneer Fund Is at the Center of Debate on Race and Intelligence," *The Chronicle of Higher Education* 41, no. 15 (December 7, 1994): A28–29.

Original works from the postwar period include W.E.B. Du Bois, "The Negro and Imperialism" in *W.E.B. Du Bois Speaks: Speeches and Addresses 1920–*

1963, ed. Phillip S. Foner (New York: Pathfinder Press, 1970); W. C. George, *The Biology of the Race Problem,* prepared by commission of the Governor of Alabama (New York: National Putnam Letters Committee, 1962); Charleton Putnam, *Race and Reason: A Yankee View* (Washington, D.C.: Public Affairs Press, 1961); and F.C.J. McGurk, "A Scientist's Report on Race Differences," *U.S. News and World Report,* September 21, 1956, a6, a2.

Scientific references on the developing views of genetic diversity are Th. Dobzhansky, *Genetics and the Origin of Species,* 2d ed. (New York: Columbia University Press, 1941); Leslie C. Dunn and Theodosius Dobzhansky, *Heredity, Race, and Society* (New York: Mentor Books, New American Library, 1946); A. S. Wiener, "Rh-Hr Blood Types in Anthropology," *Yearbook of Physical Anthropology* 1 (1945): 212–213; A. S. Wiener, "Blood Grouping Tests in Anthropology," *American Journal of Physical Anthropology* 6 (1948): 236–237; William C. Boyd, *Genetics and the Races of Man* (Boston: Little, Brown, 1950).

Modern analyses include Jared Diamond, "Race without Color," *Discover,* November 13, 1994, 82–91; L. Luca Cavalli-Sforza, Paolo Menozzi, and Alberto Piazza, *The History and Geography of Human Genes* (Princeton, N.J.: Princeton University Press, 1994); J. M. Friedman, F. J. Dill, M. Hayden, and B. McGillivray, *Genetics,* 2d ed. (Baltimore, Md.: National Medical Series for Independent Study, Williams and Wilkins, 1996), 75, table 4–5; W. Klug and M. Cummings, *Concepts of Genetics,* 5th ed. (Upper Saddle River, N.J.: Prentice Hall, 1997); R. Lewis, *Human Genetics: Concepts and Applications* (Dubuque, Iowa: Wm. C. Brown, 1997); and A. Templeton, "The Genetic and Evolutionary Significance of Human Races," in *Understanding Race and Intelligence,* ed. J. Fish (Northvale, N.J.: Jason Aronson, in press).

Chapter 10. The Race and IQ Fallacy

The following references are relevant to the modern analysis of the significance of genetic diversity in humans: F. Livington, "On the Non-Existence of Human Races," *Current Anthropology* 3, no. 3 (1962): 279–281; C. Loring Brace, "On the Race Concept," *Current Anthropology* 5, no. 4 (1964): 313–320; R. C. Lewontin and J. L. Hubby, "A Molecular Approach to the Study of Genic Heterozygosity in Natural Populations. II. Amount of Variation and Degree of Heterozygosity in Natural Populations of *Drosophila pseudoobscura,*" *Genetics* 54 (1966): 595–605; R. C. Lewontin, "The Apportionment of Human Diversity," *Evolutionary Biology* 6 (1972): 381; M. Nei and A. K. Roychoudhury, "Genetic Relationship and Evolution of Human Races," *Evolutionary Biology* 14 (1982): 1–59; Motoo Kimura, *The Neutral Theory of Molecular Evolution* (Cambridge: Cambridge University Press, 1983); M. Nei and G. Livshits, "The Genetic Relationships of Europeans, Asians, and Africans and the Origin of Modern *Homo sapiens,*" *Human Heredity* 39 (1989): 276–281; L. Luca Cavalli-Sforza, *The Genetics of Human Races* (Burlington, N.C.: Carolina Biology Readers, 1983); L. Luca Cavalli-Sforza, Paolo Menozzi, and Alberto Piazza, *The History and Geography of Human Genes* (Princeton, N.J.: Princeton University Press, 1994); J. L. Graves, "Evolutionary Biology and Human Variation: Biological

Determinism and the Mythology of Race," *Sage Race Relations Abstracts* 18, no. 3 (1993): 4–34; and J. Diamond, "Race without Color," *Discover,* November 13, 1994, 82–91.

The concepts of race and intelligence are well discussed in Carl Brigham, *A Study of American Intelligence* (Princeton, N.J.: Princeton University Press, 1923); A. Jensen, "How Much Can We Boost IQ and Scholastic Achievement?" *Harvard Educational Review* 39 (1969): 1–123; A. R. Jensen, *Educability and Group Differences* (New York: Harper and Row, 1973); Stephen Jay Gould, "The Real Error of Cyril Burt," in *The Mismeasure of Man,* revised and expanded edition (New York: W. W. Norton, 1996); S. Scarr, *Race, Social Class, and Individual Differences in IQ* (Hillsdale, N.J.: Erlbaum, 1981); R. C. Lewontin, S. Rose, and L. J. Kamin, *Not in Our Genes: Biology, Ideology, and Human Nature* (New York: Pantheon Books, 1984); D. Paul, "Textbook Treatments of the Genetics of Intelligence," *Quarterly Review of Biology* 60, no. 3 (1985): 317–326; P. A. Vernon, ed., *Biological Approaches to the Study of Human Intelligence* (Norwood, N.J.: Ablex, 1993); A. R. Jensen, *The g Factor: The Science of Mental Ability* (Westport, Conn.: Praeger, 1998).

The flexibility of the brain and concepts and measures of "intelligence" are discussed in L. L. Thurstone, *The Vectors of the Mind* (Chicago: University of Chicago Press, 1924); L. L. Thurstone, *Multiple Factor Analysis* (Chicago: University of Chicago Press, 1947); R. C. Tryon, "Individual Differences," in *Comparative Psychology,* ed. F. A. Moss, 330–365 (Englewood Cliffs, N.J.: Prentice-Hall, 1942); D. Wechsler, *The Measurement and Appraisal of Adult Intelligence* (Baltimore, Md.: Williams and Wilkins, 1958); J. Hirsch, ed., *Behavior-Genetic Analysis* (New York: McGraw-Hill, 1967); Th. Dobzhansky and A. Montagu, "Natural Selection and the Mental Capacity of Mankind," in *Race and IQ,* ed. Ashley Montagu, 104–111 (Oxford: Oxford University Press, 1975); Marian C. Diamond, *Enriching Heredity: The Impact of the Environment on the Anatomy of the Brain* (New York: Free Press, 1988); Jean Khalfa, ed., *What Is Intelligence?* (Cambridge: Cambridge University Press, 1994); Dale Purves, *Neural Activity and the Growth of the Brain* (Cambridge: Cambridge University Press, 1994): G. Schlaug, L. Jancke, Y. Huang, and H. Steinmetz, "In Vivo Evidence of Structural Brain Asymmetry in Musicians," *Science* 267 (1995): 699–701; R. Gur, L. H. Mozely, and E. Gur, "Sex Differences in Regional Cerebral Glucose Metabolism during a Resting State," *Science* 267 (1995): 528–531; and R. J. Greenspan, "Understanding the Genetic Construction of Behavior," *Scientific American,* April 1995, 72–78.

References that examine the difficulty of calculating the heritability of intelligence are M. Feldman and R. C. Lewontin, "The Heritability Hangup," *Science* 190 (1974): 1163–1168; D. Layzer, "Heritability of IQ Scores: Science or Numerology?" *Science* 183 (1974): 1259–1266; U. Bronfenbrenner, "Nature with Nurture," in *Race and IQ,* ed. Ashley Montagu, 114–144 (Oxford: Oxford University Press, 1975); R. C. Lewontin, "Genetic Aspects of Intelligence," *Annual Review of Genetics* 9 (1975): 387–405; O. Kempthome, "Logical, Epistemological, and Statistical Aspects of Nature-Nurture Data Interpretation," *Biometrics* 34 (1978): 1–23; B. Devlin, S. F. Fienberg, D. P.

Resnick, and K. Roeder, "Wringing the Bell Curve: A Cautionary Tale about the Relationships among Race, Genes, and IQ," *Chance* 8, no. 3 (1995): 27–36; J. L. Graves and T. Place, "Race and IQ Revisited: Figures Never Lie, but Often Liars Figure," *Sage Race Relations Abstracts* 20, no. 2 (1995): 4–50; and R. Plomin, "The Genetic Basis of Complex Behaviors," *Science* 264 (1994): 1733–1739.

The following sources discuss bias in aspects of standardized testing: P. Rosser, *The SAT Gender Gap: Identifying the Causes* (New York: Center for Women Policy Studies, 1989); C. M. Steele and J. Aronson, "Stereotype Threat and the Intellectual Test Performance of African Americans," *Journal of Personality and Social Psychology* 69, no. 5 (1995): 797–811; and M. Lovaglia, Jeffrey W. Lucas, Jeffrey A. Houser, Shane Thye, and Barry Markovsky, "Status Processes and Mental Ability Test Scores," *American Journal of Sociology* 104, no. 1 (1998): 195–228.

Discussions of institutional racism and environmental justice can be found in L. Knowles and K. Prewitt, eds., *Institutional Racism in America* (Englewood Cliffs, N.J.: Prentice-Hall, 1969); Richard Kluger, *Simple Justice: The History of Brown v. Board of Education and Black America's Struggle for Equality* (New York: Vintage Books, 1977); Willie Pearson Jr. and Kenneth Bechtel, eds., *Blacks, Science, and American Education* (New Brunswick, N.J.: Rutgers University Press, 1989); and Bunyan Bryant, and Paul Mohai, eds., *Race and the Incidence of Environmental Hazards: A Time for Discourse* (Boulder, Colo.: Westview Press, 1992).

Chapter 11. The Race and Disease Fallacy

Additional information concerning the history of African Americans and medicine can be found in James Jones, *Bad Blood: The Tuskegee Syphilis Experiment* (New York: The Free Press, 1981); and Keith Wailoo, *Drawing Blood: Technology and Disease Identity in Twentieth-Century America* (Baltimore: Johns Hopkins University Press, 1997).

The following general references support this chapter: L. Luca Cavalli-Sforza, *The Genetics of Human Races* (Burlington, N.C.: Carolina Biology Readers, 1983); Gordon Edlin, *Genetic Principles: Human and Social Consequences* (Boston: Jones and Bartlett Publishers, 1984); Graham R. Serjeant, *Sickle Cell Disease* (New York: Oxford University Press, 1985); Arun K. Roychoudhury and Masatoshi Nei, *Human Polymorphic Genes: World Distribution* (New York: Oxford University Press, 1988); Anthony P. Polednak, *Racial and Ethnic Differences in Disease* (New York: Oxford University Press, 1989); Randolph M. Neese and George C. Williams, *Why We Get Sick: The New Science of Darwinian Medicine* (New York: Times Books, 1994): Kenneth Weiss, *Genetic Variation and Human Disease: Principles and Evolutionary Approaches* (Cambridge: Cambridge University Press, 1995); J. M. Friedman, F. J. Dill, M. Hayden, and B. McGillivray, *Genetics,* 2d ed. (Baltimore, Md.: National Medical Series for Independent Study, Williams and Wilkins, 1996); W. Klug and M. Cummings, "Genetics and Cancer," in *Genetics,* 5th ed. (Upper

Saddle River, N.J.: Prentice Hall, 1997); R. Lewis, *Human Genetics: Causes and Applications,* 2d ed. (Dubuque, Iowa: W. C. Brown, 1997).

The evolutionary theory of aging is best summarized by Michael R. Rose, *The Evolutionary Biology of Aging* (New York: Oxford University Press, 1991). For more recent reviews, see M. R. Rose and C. Finch, "Hormones and the Physiological Architecture of Life History Evolution," *Quarterly Review of Biology* 70, no. 1 (1995): 1–52; M. Jazwinski, "Longevity, Genes, and Aging," *Science* 273 (1996): 54–59. I examine reductionism in the biological aging paradigms in J. L. Graves, "General Theories of Aging: Unification and Synthesis," in *Principles of Neural Aging,* ed. Sergio U. Dani, Akira Hori, and Gerhard F. Walter, 35–55 (Amsterdam: Elsevier Press, 1997).

Citations from the primary literature concerning AIDS include the following: C. Quillent, "HIV-1 Resistance Phenotype Conferred by Combination of Two Separate Inherited Mutations of CCR5 Gene," *The Lancet* 351, no. 9095 (1997): 14; N. L. Michael, L. G. Louie, and H. W. Sheppard, "The Role of CCR5 and CCR2 Polymorphisms in HIV-1 Transmission and Disease Progression," *Nature Medicine* 3, no. 10 (1997): 1160; S. O'Brien and M. Dean, "In Search of AIDS Resistance Genes," *Scientific American,* September 1997, 44–53; J. J. Martinson, N. H. Chapman, and J. B. Clegg, "Global Distribution of the CCR5 32 Basepair Deletion," *Nature Genetics* 16, no. 1 (1997): 100; and J. Caldwell and P. Caldwell, "The African AIDS Epidemic," *Scientific American,* March 1996, 62–69.

Citations on cancer include P. A. Wingo, S. Bolden, T. Tong, S. Parker, L. Martin, and C. Heath, "Cancer Statistics for African Americans," *Cancer Journal for Clinicians* 46 (1996): 113–125; C. Baquet, J. W. Horm, T. Gibbs, and P. Greenwald, "Socioeconomic Factors and Cancer Incidence among Blacks and Whites," *Journal of the National Cancer Institute* 83, no. 8 (1991): 551–556; M. Roach, J. Krall, J. W. Kellar, C. A. Perez, W. T. Sause, R.L.S. Doggett, M. Rotman, H. Russ, M. V. Pilepich, S. O. Asbell, and W. Shipley, "The Prognostic Significance of Race and Survival from Prostate Cancer Based on Patients Irradiated on Radiation Therapy Oncology Group Protocols (1976–1985)," *International Journal of Radiation Oncology, Biology, Physics* 24 (1992): 441–449; A. Whittemore et al., "Prostate Cancer in Relation to Diet, Physical Activity, and Body Size in Blacks, Whites, and Asians in the United States and Canada," *Journal of the National Cancer Institute* 87, no. 9 (1995): 652–661; S. A. Devgan, B. E. Henderson, M. C. Yu, C. Y. Shi, M. C. Pike, R. K. Ross, and J. K. Reichardt, "Genetic Variation of 3 beta-Hydroxysteroid Dehydrogenase Type II in Three Racial/Ethnic Groups: Implications for Prostate Cancer Risk," *Prostate* 33, no. 1 (1997): 9–12; and B. A. Miller et al., eds., *Racial/Ethnic Patterns of Cancer in the United States, 1988–1992* (Bethesda, Md.: National Cancer Institute, 1996).

Citations on hypertension and CVD include P. P. Moll, E. Harburg, T. L. Burns, M. A. Schork, and F. Orgoren, "Heredity, Stress, and Blood Pressure, a Family Set Approach: The Detroit Project Revisited," *Journal of Chronic Diseases* 36 (1983): 317; C. L. Broman, "Social Mobility and Hypertension among Blacks," *Journal of Behavioral Medicine* 12, no. 2 (1989): 123; D. Calhoun,

"Hypertension in Blacks: Socioeconomic Stress and Sympathetic Nervous System Activity," *American Journal of Medical Sciences* 304, no. 5 (1992): 306; R. S. Cooper and C. N. Rotimi, "Hypertension in Populations of West African Origin: Is There a Genetic Predisposition?" *Journal of Hypertension* 12 (1994): 215–227; K. Light, "Job Status and High-Effort Coping Influence Work Blood Pressure in Women and Blacks," *Hypertension* 25, no. 4 (1995): 1; J. S. Kaufman, R. A. Durazo-Arvizu, C. N. Rotimi, D. L. McGee, and R. S. Cooper, "Obesity and Hypertension Prevalence in Populations of African Origin: Results from the International Collaborative Study on Hypertension in Blacks," *Epidemiology* 7, no. 4 (1996): 398–405; J. S. Kaufman, C. N. Rotimi, W. R. Brieger, M. A. Oladokun, S. Kadiri, B. O. Osotimehin, and R. S. Cooper, "The Mortality Risk Associated with Hypertension: Preliminary Reports of a Prospective Study in Rural Nigeria," *Journal of Human Hypertension* 10 (1996): 461–464; W. B. Neser, J. Thomas, K. Semenya, D. J. Thomas, and R. F. Gillum, "Obesity and Hypertension in a Longitudinal Study of Black Physicians: The Meharry Cohort Study," *Journal of Chronic Diseases* 39 (1986): 105–113; and M. A. Winkleby, H. C. Kraemer, D. Ahn, and A. N. Varady, "Ethnic and Socioeconomic Differences in Cardiovascular Disease Risk Factors," *JAMA* 280, no. 4 (1998): 356–362.

Conclusion

General references are Ashley Montagu, *Man's Most Dangerous Myth: The Fallacy of Race,* 5th ed. (New York: Oxford University Press, 1974); J. Philippe Rushton, *Race, Evolution, and Behavior: A Life History Perspective* (New Brunswick, N.J.: Transaction Publishers, 1995); and Michael Levin, *Why Race Matters: Race Differences and What They Mean* (Westport, Conn.: Praeger, 1997).

On the flexibility of brain development, see Marian C. Diamond, *Enriching Heredity: The Impact of the Environment on the Anatomy of the Brain* (New York: The Free Press, 1988); and Dale Purves, *Neural Activity and the Growth of the Brain* (Cambridge: Cambridge University Press, 1994).

Appendix B

R. C. Lewontin and J. L. Hubby, "A Molecular Approach to the Study of Genetic Heterozygosity in Natural Populations. II. Amount of Variation and Degree of Heterozygosity in Natural Populations of *Drosophila pseudoobscura,*" *Genetics* 74 (1966): 595–609; M. Nei and A. K. Roychoudhury, "Gene Differences between Caucasian, Negro, and Japanese Populations," *Science* 177 (1972): 434–435; and L. Luca Cavalli-Sforza, Paolo Menozzi, and Alberto Piazza, *The History and Geography of Human Genes* (Princeton, N.J.: Princeton University Press, 1994), 25–27.

INDEX

About the Author

Joseph L. Graves Jr. is the author of *The Race Myth* and numerous book chapters and papers on biological concepts of race. He has held professional appointments in evolutionary biology and African American studies at the University of California, Irvine, Arizona State University, Fairleigh Dickinson University, and is currently Dean of University Studies and Professor of Biological Sciences at North Carolina A&T State University.

CPSIA information can be obtained
at www.ICGtesting.com
Printed in the USA
LVHW110712150921
697820LV00012B/78